S0-ASE-375

THE GENIE IN THE BOTTLE

THE GENIE IN THE BOTTLE

64 All New Commentaries on the Fascinating Chemistry of Everyday Life

DR. JOE SCHWARCZ
Director
McGill University Office for Chemistry and Society

W. H. Freeman and Company
New York

The publication of *The Genie in the Bottle* has been generously supported by the Canada Council, the Ontario Arts Council, and the Government of Canada through the Book Publishing Industry Development Program.

Cover design by Guylaine Regimbald – SOLO DESIGN.
Cover Illustration by Paul Cozzolino/Laughing Stock.
Copyedited by Mary Williams.
Interior design by Yolande Martel.
Interior cartoons by Brian Gable.
Author photo by Tony Laurinaitis.

Cataloging-in-publication data available from the Library of Congress

Published in the United States in 2001 by W. H. Freeman and Company,
41 Madison Avenue, New York, New York 10010.

Printed in the United States of America

First US printing 2001.

CONTENTS

FOOD MATTERS

CHEMISTRY HERE, THERE, EVERYWHERE

PREFACE

It was a dark and stormy night. Really. There was a knock at the door. The well-dressed gentleman on my doorstep introduced himself and proceeded to ask me a rhetorical question: "Are you interested in good health?" For a moment I pondered putting an end to the encounter by saying, "No, I would rather be cold, hungry, and sick," but I thought better of it. Why not let him have his say? I invited him in.

"I notice you have a tap in the kitchen," he began, quickly demonstrating his keen powers of observation. I admitted that we had, indeed, opted for a house with indoor plumbing, despite the health benefits we might have attained by carrying buckets of well water during the Canadian winter. "You don't actually drink that water, do you?" he went on. As if admitting to a crime, I replied that not only did we drink the tap water, but we also gave it to the cat. This seemed to cause the man grave concern: "It has chemicals in it, you know." I think he was a little taken aback that this bombshell didn't immediately cause me to clutch my throat; he evidently decided that heavier artillery was needed. "Invisible chemicals," he explained. By now I had a pretty good idea where he was going with this, but the time was not yet ripe for a lecture on why "chemical" isn't a four-letter word.

"Would you like to see those invisible chemicals?" he asked. Before I had a chance to ask how anything invisible could be seen, he began unpacking some equipment from his briefcase. It looked impressive. It turned out to be some sort of electrical device fitted with a pair of metal rods that looked like electrodes. Next he asked for a glass of water from my tap. He sniffed it, and, apparently convinced that the liquid was sufficiently toxic, proceeded to immerse the electrodes in it. Then, with a cry of "Watch this!" he plugged the device into a wall socket. Within thirty seconds the water started to turn cloudy, and within a minute it had formed a repulsive yellow scum. "You see!" the man cried triumphantly, implying that by passing an electrical current through the water he had scared those nasty chemicals out of solution. The toxins had been comfortably dissolved, it seems, until the fear of being electrocuted prompted a mass exodus.

Then came the coup de grâce. He extracted a water filter from his bag and attached it to the tap. He proceeded to subject the filtered water to the same kind of electrocution as had been experienced by my "toxin-laden" tap water, but this time the results were dramatically different. There was no yellow sludge, because those invisible, noxious chemicals had been "filtered out." Surely, the couple of hundred dollars this miraculous filter cost was a small price to pay for my family's health. But if I was still unconvinced, the salesman told me, he had lots of documentation to support his claims. Out came the newspaper clippings about the various dangers that lurked in tap water, including expert testimony on how chlorine had been used as a poison gas during World War I. Again he reached into his bag. I waited for him to pull out a gas mask — I'd been wondering how he had dared to confront a lethal water tap without suitable protection in the first place. But no, instead of a gas mask, he grabbed a bottle of ortho-tolidine. He informed me that this substance would reveal the presence of chlorine in water by turning yellow. Sure enough, my tap water contained chlorine.

Now the man asked me to place a couple of fingers in a fresh glass of tap water and wait a few minutes. He tested the water again with ortho-tolidine, and this time there was no telltale yellow color. The toxic chlorine, he insisted, had been absorbed into my body. Exactly the same process occurred, I was told, every time I took a shower. No need to give up showers, though: the filtered water had no chlorine residue, and he had a filter that would fit any shower. With this dramatic demo, my lesson in toxicology and chemistry came to an end.

It wasn't easy, but I bit my tongue and made it all the way through. I didn't even react when the salesman talked about "soaring cancer rates," "bodies overburdened with toxins," and "scientists brewing up deadly chemical mixes." I resisted pointing out that the average life expectancy lengthens every year and that while some cancers are increasing, others are declining. I didn't even mention that the introduction of water chlorination was probably the greatest public health advance in history. But now it was my turn. Time for me to give a little chemistry lesson.

I began by picking up the glass of yellowed, scummy tap water, the one in which those nasty chemicals were no longer invisible, and raising it to my lips. Before the salesman had a chance to stop what must have seemed like a suicide attempt, I downed the contents. At this point the poor man's face turned the color of the liquid in the glass. He must have thought I was mad. But I knew that I wasn't taking any risks; I'd figured out what was happening. The yellow sludge wasn't coming from invisible chemicals that had been jolted out of solution — it was coming from one of the electrodes. Electrolysis is a classic chemical experiment in which two electrodes are immersed in water and a current is passed between them. This causes water to break down into oxygen and hydrogen. But if one of the electrodes is made of iron, it reacts with water to form a precipitate of yellow iron hydroxide — or rust.

So, all I was doing was drinking a little rust. Just a form of iron supplement, I explained to the incredulous salesman. I decided to punctuate my little performance by taking his glass of filtered water, adding a few grains of salt, and subjecting it to a current. Within seconds the familiar yellow scum formed. The salesman watched in awe. What kind of a magician was I? He was confused. I explained to him that water only conducts electricity when it has ions dissolved in it, and his filter had removed these ions. Therefore, no scum. But when I added a little salt, electricity flowed through the water and allowed the iron electrode to rust. To prove my point, I replaced the iron electrode with an aluminum one and invited him to torture my tap water with his apparatus once again. Since this time he used no iron electrode, there was no scum.

Next we tackled the chlorine problem. I drew two glasses of tap water and placed them on the table. I inserted a couple of fingers in one and asked the salesman to hold the other. A few minutes later we tested each for chlorine content. Neither glass had any. Chlorine, I explained, evaporates. It goes into the air, not into the skin. I wasn't sure how effective my arguments and demonstrations had been. The salesman pointed out that the scum had formed with tap water and not with filtered water, so the filter had done something. I couldn't argue with that logic.

Certainly, this was not the only occasion that I'd found myself listening to curious chemical stories and bewildering claims. The business of bringing science to the public through books, newspapers, radio, and television tends to prompt requests for consultations. Over the last twenty years or so, an assortment of entrepreneurs has visited me at home or at my office either to solicit my opinion on a product or to entice me into a "can't lose" business venture based on some miraculous cure-all. I've seen and heard everything: crystals, magnets, pyramids, countless dietary supplements, convoluted weight-reduction schemes, special oils, oxygenated liquids, deoxygenated liquids, odor removers, odor producers, exotic juices, ionizing bracelets, herbal concoctions, antioxidants of every description, parasite killers, therapeutic glasses, foot deodorants, water magnetizers, blankets that heal, and charcoal-laden underwear that counters the effects of bean consumption.

By and large, the people I've met, and continue to meet, are well meaning and not out to defraud others. But they do share an unrealistic and overly simplistic view of the way the world works. They bandy terms like "toxins," "chemicals," and "poison" about recklessly while misguidedly revering "natural" substances. Most possess only the vaguest understanding of molecules, chemical reactions, and research methods. They have little appreciation of the power of the placebo or the confusion that can be created by undue reliance on anecdotal evidence. True, science does not have all the answers, and scientists do make mistakes, but sticking to the scientific method is still our best shot at progress.

The purpose of this work — and of *Radar, Hula Hoops and Playful Pigs*, my previous attempt at demystifying science — is to provide a few scientific glimpses into the workings of our complex world. My hope is that by offering explanations for a variety of common phenomena I can help the reader develop a feel for how the scientific method functions, and at the same time, lay down a solid foundation for critical thinking.

That, of course, was just what I had in mind as I met with the water-filter salesman. I could lead him to water, as it were, but could I make him drink? My demos and explanations may have had some effect, because when I offered him a cup of coffee, he happily sipped it, despite the fact that it had been brewed with tap water. It was then that I decided I had tormented the poor soul enough and that I should reward him for sitting through my chemistry lecture. He just about fell off his chair when I said I would buy a filter. Of course, my decision had nothing to do with his irrelevant demonstrations. I had been contemplating purchasing a filter, anyway. These devices do remove a number of undesirable substances that escape municipal treatment — trihalomethanes, for one. While chlorine unquestionably saves millions of lives, we do pay a price for using it. Chlorine reacts with some dissolved organic compounds to produce trihalomethanes, which are carcinogenic. Activated carbon filters remove these, as well as a variety of other pollutants. While the risk of drinking tap water is very small compared with other risks we face, it is one that we can easily reduce by using a good filter. In any case, water tastes better when it is free of chlorine residues.

So, I wrote out a check for the filter, gave my new friend a chemistry text, and hoped that he would reap some benefit from our visit. The night, I thought, had been dark and stormy for him in more ways than one. I watched through the window as he braved the weather and headed towards the next house. He paused for a moment. I guess he needed some stress relief. The man who was so worried about the chemicals in my tap water reached into his pocket, pulled out a cigarette, and lit up.

INTRODUCTION

Of Rabbits, Elephants, and Genies

I used to have a rabbit. His name was Ether. Billed as "the Ether Bunny," he was the star of the *Magic of Chemistry* stage show. A few years ago a couple of colleagues and I devised this spectacle to get students and the public excited about chemistry. It turned out to be so much fun that we're still at it. Using a variety of chemical demonstrations, supported by slides and music, we describe the evolution of chemistry and explain its role in the development of everyday products, ranging from dyes and cosmetics to gunpowder and nylon. Some of the demos really do seem to be magical. Solutions spontaneously change color, flames mysteriously flare up, and liquids appear to disappear. But the theme of the show is that this isn't magic. The magic only exists as long as there is no explanation; as soon as one is provided, the magic evolves into science.

To distinguish science clearly from magic, we blend a few carefully selected tricks into the show. We explain how our enjoyment of magic depends on our not knowing how an effect is carried out, for if the method is revealed a spectacular illusion often changes into a simple trick. Scientists, on the other hand, revel in providing answers, so we consider that the real magic of our demonstration is that it illuminates how and why something happens.

The logo for our show depicts a rabbit emerging from a hat, holding a chemical flask. This sets up the lighthearted ending of the presentation, when we say to the audience that, despite the spectacular effects they have just witnessed, they are probably disappointed because they've come to a magic show and haven't yet seen a rabbit. Then I produce a top hat ("the natural habitat for a rabbit") and ask the audience what they would like to see come out of it. Amid the expected cries of "A rabbit!" Ether emerges triumphant — or at least he used to.

The problem with magical rabbits is that they have to be kept between performances. And while Ether was a great stage performer, his performance around the house was less than admirable. Since I didn't have the heart to keep him in a cage all the time, I let him wander. He took advantage of this by leaving his calling cards everywhere and stripping every telephone wire of its plastic insulation. When Ether eventually departed for that great rabbit warren in the sky (or perhaps elsewhere) I didn't get a new rabbit, and *The Magic of Chemistry* suffered. Something had to be done, but I was no longer willing to live with rabbit pellets under my desk. I replaced Ether with a synthetic rabbit.

This had a certain advantage: it allowed me to engage in some patter about how chemists can make anything, including rabbits — as long as they are synthetic, of course. All went well, until, during one performance, I asked the audience what they would like to see come out of the hat and I didn't receive the usual reply. A boy with well-exercised lungs yelled out, "An elephant!" The audience took delight in the little fiend's antics, and soon the auditorium reverberated with their own demands for an elephant. Poor synthetic Ether was received with derision. I vowed that this would never happen again.

Somehow I had to be ready for the next young menace who was infatuated with pachyderms. But how? Elephants were out of the question. I'd had enough problems with rabbit droppings around the house. Then I remembered reading about a chemical demo called "elephant toothpaste," whereby a couple of chemicals are combined in a flask and, almost immediately, a long, foamy strand is produced that looks, for all intents and purposes, like a toothpaste for elephants — if elephants used toothpaste, that is. Here was my solution. I would be ready for the next smart aleck. I would dig around in the hat to find the elephant but would fail to produce him. Why? Because he was busy. He was brushing his teeth, I would tell the audience. "Don't believe me? He's squeezing the toothpaste tube right now!" Then I'd concoct the toothpaste. I practiced the effect until I had it down pat.

All that I needed was an opportunity to perform the feat. It finally came, although I must admit I prompted the audience by asking what animal "that could not possibly be hidden in the hat" they would most like to see. I made a profusion of elephant toothpaste, and there was a bonus. The demonstration allowed me to fill people in on the chemistry involved.

Hydrogen peroxide decomposes in the presence of potassium iodide to yield oxygen and steam. If we add some detergent to the mix, a foam forms as the oxygen and steam emerge from the flask. It's neat.

As it turned out, this reaction was very similar to another hydrogen peroxide decomposition that we had been using in the show. A chemical genie, basically steam, spurted out of an old-fashioned bottle I had found at a flea market, leading to discussion of how chemistry is like a genie. It can do a great deal of good if you use it the right way, but if you are thoughtless, the consequences can be dire. Furthermore, it may be easy to produce a genie, but it's a lot harder to put it back into the bottle. So the elephant toothpaste and genie demos connected nicely, and each could be milked for its entertainment and educational value. And the whole escapade served yet another purpose. It provided me with a rather enchanting title and cover design for this book.

HEALTH MATTERS

Surviving the Rat Race

I think I was about twelve years old when I attended my first university lecture. No, I was not a child prodigy. And I certainly did not go willingly. My parents dragged me to McGill University to listen to a talk given by Dr. Hans Selye, who was at that time already a recognized world authority on stress. I'm not sure why my parents were bent on attending this event, but I suspect that it was because, like them, the good doctor was Hungarian. At least he had a Hungarian name — Selye was actually an Austrian who had been educated in Prague, Paris, and Rome, but Hungarians tend to insist that if you have a Hungarian name, then you're Hungarian.

Memory is a strange thing. I couldn't tell you what Dr. Selye's lecture was about, but I vividly remember one story that he told in the course of it. It was about meeting a drunk who was mildly abusive. Selye had a decision to make. He could either get into a physical confrontation with the chap, or ignore him and walk away. A fight would have elevated his blood pressure (I remember this, because as he said it he waved a blood pressure cuff around wildly) and increased his pulse rate, effects that he decided were best avoided. Then Selye described another situation, in which he was mugged and threatened on the street. This time there was no turning the other cheek — action was required. His pulse raced, his blood pressure soared, and he beat his attacker off. This, I must admit, was a touch difficult to believe, because Selye hobbled as a result (I later learned) of two hip operations.

I didn't get the point of his story at the time. In fact I didn't get it until about twenty-five years later, when I read Selye's classic work *The Stress of Life*. By then I'd discovered that Selye was probably the world's leading authority on what was being called "biological stress syndrome." Actually, he was more than an authority; he was the originator of the term. At McGill University, during the 1930s, Selye had carried out a series of experiments on rats, injecting them with a variety of toxins. While the rats manifested different reactions depending on which toxin Selye had used, they also shared a number of symptoms irrespective of the nature of the toxin. The rats' adrenal cortexes enlarged, their spleens and thymus glands shrank, and bleeding ulcers developed in their guts. In other words, they reacted to the stress. Selye then went on to show that he could produce the same reactions by subjecting the rats to demanding physical or psychological conditions. Stress, alone, was capable of triggering chemical reactions in the body.

It didn't take Selye long to uncover exactly what was going on. Under stress, the adrenal gland pumped out adrenaline and cortisol, which then caused the physical symptoms. And this happened not only in rats but also in humans. Stress, it seemed, could raise our blood pressure, make us sweat, and force our hearts to beat faster. If we had underlying heart disease, it could even kill us. But "could" was Selye's key word. Stress didn't have to have negative effects on our bodies — not if we could adapt ourselves to it. The way in which we handled an adverse situation, not the situation itself, was critical. And this is where

the story of the drunk comes in. As Selye maintained, we can choose how we will react to a stressful situation: get angry and provoke a potentially dangerous physiological response; or simply walk away. We often find ourselves confronted with such a choice. You find a parking ticket on your windshield. Do you rant and rave and then pay the ticket, or do you calmly accept the fact that you were negligent and then pay the ticket? In either case the financial penalty is the same, but the health penalty may be quite different.

Of course, we don't always have a choice. That's why our bodies have evolved the ability to secrete adrenaline and cortisol. Sometimes we need a sudden burst of energy, a boost to the heart's pumping capacity. Sometimes we have to flee; and sometimes we're obliged to fight, as Selye was when he faced the mugger. Decide what is worth fighting for and what is not, was Selye's message, because it may be a matter of life or death.

Selye's basic assertion was that inappropriate negative emotions can be physically destructive. If that is so, then what can positive emotions do? In 1964 that very question popped into the mind of Norman Cousins. The well-known writer and editor had developed a form of arthritis that attacks the body's connective tissue. Ankylosing spondylitis is a terrible disease that, as it progresses, usually immobilizes its victims by welding together joints, particularly those in the spine. Could positive responses to emotion, like laughter, be of any help in battling his ailment, Cousins wondered? In an effort to find out, he decided to undergo "laughter therapy." He rented movies starring the Marx Brothers and Abbott and Costello (yes, some people find them funny) and began to laugh his way back to health. Within eight days Cousins noticed some improvement. Four months later he was back at work on his way to conquering the disease. Cousins recounted his remarkable escapade in his book *Anatomy of an Illness.*

Was Norman Cousins really cured by laughter, or was he just one of the lucky few who recover from ankylosing spondylitis? That's a difficult question to answer, but many no doubt tried to follow in his footsteps and laugh themselves to health, only to succumb to their disease. And these people didn't write books about their experiences. Cousins's self-cure may be questionable, but his contribution to science is undeniable. He sensitized the scientific community to our need to study body-mind relationships seriously, and his efforts have led to some fascinating observations.

A study conducted at Stanford University Medical Center found that breast cancer patients enrolled in support groups where they shared feelings, learned stress-reduction techniques, and always had access to a good listener, were less depressed, experienced less pain, and enjoyed a more positive outlook. They also survived twice as long. If these results had been obtained with a new medication, pharmaceutical companies would have revved up their publicity machines. Yet, even so, the findings have made their mark. Wonderful self-help groups, like Gilda's Club (named after comedienne Gilda Radner, who died of cancer), have sprung up, easing the burden of cancer patients.

Researchers at Texas A&M University discovered that mood, blood pressure, and surgery recovery time can be influenced by art — but not just any kind of art. Patients who had Picasso reproductions in their rooms fared worse than those with blank walls, while some of those who gazed at Monet's water lilies recovered more quickly. I think Hans Selye must have loved beautiful paintings, too. After all, he was himself an artist of sorts. He carved part of the cortisol molecule into the cement outside his window when he was living on Milton Street near McGill. It's still there — a silent testimonial to the man my parents dragged me to see on that stressful day so long ago.

Thalidomide: A Bitter Lesson

On August 2, 1962 Dr. Frances Kelsey, a McGill University graduate, stood proudly as President Kennedy hung the President's Award for Distinguished Federal Civilian Service around her neck. What had Dr. Kelsey done to deserve this honor? She had spared thousands of children from being born with disfigurements ranging from seal-like flippers instead of hands or arms to distorted heads with no ears. Kelsey, working for the Food and Drug Administration, almost single-handedly prevented the U.S. sale of a drug that in other countries would ultimately cause at least eight thousand children to be born with severe birth defects. The drug was thalidomide.

In 1957 a brand new medication was introduced in Europe, and its manufacturers made some amazing claims. The wonder drug, they said, would cure insomnia with none of the side effects or dangers of barbiturates. It was so safe that you couldn't even commit suicide by taking a handful of the pills. And, to the delight of many pregnant women, the tranquilizer seemed ideal for the treatment of morning sickness. In Germany,

Great Britain, and Canada thousands of pregnant women took thalidomide without a second thought. But not in the United States. There, the William S. Merrel Company, thalidomide's American licensee, ran into a feisty roadblock in the person of Frances Kelsey. Dr. Kelsey was the FDA officer responsible for reviewing Merrel's drug-marketing application. As she read through the application and the mountains of submitted data she became bothered by the fact that the drug's sleep-inducing effect, which was so evident in people, had not been evident in many of the test animals. Yet the safety data was virtually all based on animal studies. Questions rose in Kelsey's mind. If animals reacted differently to thalidomide, then was the drug behaving differently in animals? And, if so, were the animal safety studies relevant?

There was another troublesome point. The writer of a letter to a British medical journal had claimed that some patients on thalidomide experienced a tingling sensation in the fingers. Kelsey wondered what this could mean in a pregnant woman. Could the fetus perhaps be affected in some way? No, Merrel insisted, they had studied the drug's effects on pregnant rats and even pregnant women. Indeed, they had. But the problem was that they didn't study the pregnant women in their first trimesters. And that was exactly when, as it later turned out, thalidomide wreaked havoc with the developing fetus. The rats were not susceptible to thalidomide's effects; the human liver, unlike that of a rat, produces an enzyme that converts thalidomide into its dangerous form. Kelsey had no knowledge of this, but she was concerned enough about the documentation to insist on more data, which irritated Merrel officials insisted was unnecessary. As the legal and technical wrangling between Kelsey and Merrel dragged on, a letter written by an Australian obstetrician and published in the prestigious British medical journal *The Lancet* rendered the whole question moot.

The obstetrician, Dr. William McBride, had noted that in his practice an unusual number of babies were being born with deformities. These deformities were unlike anything he'd ever seen. The babies suffered from phocomelia, a terrible condition whose name derives from the Greek words for *limb* and *seal*. It was a very apt name, because the newborns had little flippers instead of limbs. Checking his records McBride realized that the tragedy had only befallen babies whose mothers had been prescribed thalidomide for morning sickness. By the time his note appeared in *The Lancet* some eight thousand children had been similarly affected in Europe and Canada, and no one had made the connection to thalidomide. But now the cat was out of the bag. The drug was quickly removed from the market and Merrel withdrew its drug application. Frances Kelsey's resolve in the face of pressure and insults from Merrel had saved thousands of American children from disfigurement.

William McBride, justifiably, became an Australian hero. All of a sudden he was a celebrity to whom people looked for advice on all kinds of issues. He became Man of the Year, Father of the Year, and the head of his own research institution. But in 1987 the walls of that institution came tumbling down on McBride. After so successfully establishing the birth-defect-inducing, or teratogenic, effect of thalidomide, McBride launched a research program to examine the teratogenic potential of other medications. He decided to investigate scopolamine, because it and various related compounds were being touted as effective medications for morning sickness. McBride's scopolamine tests on rabbits produced deformed fetuses, and the doctor published a paper with his coresearchers warning of another potential disaster. It appeared that the Australian knight in shining armor had struck down another villain.

But this new revelation shocked scientists — among them the coauthors of McBride's paper. They maintained that McBride had not actually consulted them when writing the paper — and for a very good reason. They would have objected to the findings it contained. McBride had altered the data to implicate scopolamine as a teratogen. An alert medical journalist exposed the affair, which resulted in the longest and most expensive professional disciplinary hearing in world history. McBride was found guilty of scientific fraud. He claimed that at worst he was guilty of sloppy science, and that he had been made a scapegoat by the multinational drug companies, whose officials were worried about exposure, and by his medical colleagues, who were jealous of his fame. Evidently, the fame that the thalidomide revelation had conferred upon Dr. McBride was not adequate to satisfy his ego. He wanted more, and he shoved his principles aside to get it.

Some good did come of this tragic affair, however. In 1992 the U.S. Congress introduced legislation requiring far more rigorous testing of drugs before they go on the market. Medications must now be tested in at least two species of pregnant animals. Obviously, though, no matter what extensive testing may show, pregnant women should only take medication when it's absolutely necessary. Many substances cross through the placenta from the mother's bloodstream into the baby's.

We often hold up thalidomide as the ultimate example of what may occur when we fail to test drugs rigorously enough. Advocates of "natural" medicine use the thalidomide saga in their campaign to heap muck on mainstream, scientific medicine, forgetting that numerous natural substances can also have serious toxic effects. Thalidomide has become the villain. Too bad. Recent research indicates that it could be an effective treatment for a variety of conditions.

Since thalidomide can inhibit the growth of blood vessels, it is called an angiogenesis inhibitor. This is why it is so dangerous in the first trimester of a pregnancy. The blood vessels needed for limb development just do not grow. Obviously, this is undesirable; but for doctors dealing with a tumor that must develop a supply of blood vessels in order to grow, thalidomide might be just the drug they need. Similarly, in one type of macular degeneration, a terrible eye disease characterized by uncontrolled blood-vessel growth and bleeding behind the retina, thalidomide may prove to be very helpful. Then there is leprosy. One of this disease's terrible complications is the appearance of painful lesions that appear on the arms, face, and legs. Thalidomide is an effective treatment. It may even play a role in the treatment of arthritis, lupus, Crohn's disease, and multiple sclerosis, all of which are characterized by an increase in the blood of tumor necrosis factor alpha (TNF), a chemical produced by the white blood cells. TNF may be responsible for some of the symptoms of these conditions, and thalidomide inhibits its production. One day, perhaps, we'll see some world leader bestowing a medal upon the researcher responsible for establishing the healing abilities of thalidomide.

The Dark Side of the Sun

Do you know what happens to orphaned baby elephants in Kenya? Their ears, which normally look like huge flat fans, lose their rigidity and the tips fold over. This happens because the proteins that form the molecular framework of the ears are degraded by the sun. Normally, baby elephants walk underneath adult elephants and are shaded from the sun, but the orphans have to fend for themselves. So in elephant orphanages — and, yes, there are such things in Kenya — workers apply large amounts of sunscreen to the babies' ears. If you would like to picture an even more bizarre scene, imagine those workers spreading blankets over the backs of the little pachyderms to protect them from sunburn. Is there a message for us here somewhere? Yes, there is: even if we don't have big floppy ears or make a habit of wandering around the Kenyan countryside we still need sun protection. Just ask any dermatologist. Or, better yet, watch one. I did just that when an old high school friend came to town.

We had arranged to meet for lunch on the McGill University campus. My friend the dermatologist was waiting for me when I appeared. He was sitting on a bench, in the shade, sporting a wide-brimmed hat. A tube of sunblock was sticking out of his shirt pocket. That's because he knows firsthand what the sun can do. He now sees ten times as many patients with skin cancer as he did when he first started practicing in the 1970s. The girls we went to school with, at least the ones who celebrated the arrival of spring by sliding sun reflectors under their chins, are now his patients. The lucky ones exhibit the classic signs of "photoaging" and are hounding him for the latest antiwrinkle concoction or liver-spot eraser. The less fortunate ones are having basal cell or squamous cell cancers removed.

The sun reflector has, mercifully, gone the way of the leisure suit. But its effects linger. Skin cancer, like most other cancers, has a long latency period. Many people have yet to pay the price for their youthful follies. Who would have guessed? The sun was supposed to be good for us. In 1903 Icelander Niels Finsen won the Nobel Prize for medicine; he had developed a sunlight therapy to combat infectious diseases. Sunbathing became a popular treatment for tuberculosis, Hodgkin's disease, syphilis, and festering wounds. Then we discovered that jaundiced babies respond to sun exposure. But we were really sold on the sun when we learned that it triggered the formation of vitamin D in the skin. Scientists told us that all those cases of childhood rickets that occurred during the European Industrial Revolution were attributable to the clouds of smoke that belched from factory chimneys and obliterated the sun. Since we didn't want bowlegged children, we made sure they got plenty of sun. In the 1930s *The Ladies Home Journal* printed an expert's recommendation that mothers remove their babies' bonnets and uncover their feet and legs to expose them to the healthy sunshine. Coco Chanel made tanned skin a sign of affluence, and Coppertone ads warned us not to be a "paleface." How radically different things are today! We send our children off to camp clutching giant bottles of sunblock, we dress them as though their clothes were body armor against the sun, and we threaten them with never seeing another Harry Potter book if they dare to even think about taking off their hats outdoors.

I remember when, in the 1960s, Dr. Albert Kligman became the first to voice misgivings about suntanning. Kligman is arguably the most famous dermatologist in the world. When he attends a medical convention he is treated like a superstar. Some physicians seem motivated to genuflect in appreciation of Kligman's pioneering work linking skin cancer and wrinkles to sun exposure. Kligman told us something we didn't want to hear: ultraviolet light can do terrible things to skin, because that light is energetic enough to break some of the bonds that hold together the atoms in molecules, particularly in collagen and elastin, the proteins responsible for keeping the skin taut and youthful. These molecules form the molecular scaffolding on which our skin is stretched, but bombardment by high-energy ultraviolet light breaks some of the rungs, and the scaffolding collapses. The result is a leathery, wrinkled look. Anyone doubting the ability of ultraviolet light to wreak such destruction need only compare the texture of the skin on their face to that of the skin on the other end of the anatomy, the part that we rarely expose to sunshine.

But the haggard look isn't even the worst of it. There's more. Excessive sun exposure can cause eye cataracts and lead to impaired immune-system activity. That's what the sores that appear on your lips after a fun-filled vacation in the sun are all about. The herpes simplex virus will lie dormant in the body until the immune system falters, allowing the virus to multiply and produce those unsightly lesions. Sun worshippers who take certain medications are vulnerable to yet another hazard. Their skin may become even more sensitive to sunlight, and they will be tormented by burns, rashes, and skin eruptions. Some common antibiotics (tetracycline), blood pressure pills (hydrochlorothiazide), and nonsteroidal anti-inflammatory drugs (naproxen), fall into this category, as does the herbal antidepressant St. John's wort.

Without question, though, the greatest danger posed by sun exposure is skin cancer. Ultraviolet light can either disrupt DNA molecules directly, or it can generate free radicals, which then go on to attack DNA. This is not a new revelation. We have long known that far more California beach bunnies contract cancer than Quebec nuns. And we have long been hearing the advice, so eloquently expressed by the Australian authorities, to "slip, slap, and slop": slip on a T-shirt, slap on a hat, and slop on the sunscreen. Unfortunately, many of us continue to ignore it.

You don't have to be an Australian lifeguard to suffer negative reactions to the sun. In North America our leisure time has increased and our standard of living has improved, but the incidence of skin cancer has risen, as well. Some scientists believe that intense, short-term sun exposure — the kind we North Americans will receive on a vacation — is particularly dangerous. Then, to top it all off, we have to worry about our diminishing ozone layer. Ozone filters out the most damaging UV rays, or at least it used to. Now we are concerned that the destruction of ozone, mostly due to freons used in air conditioning and refrigeration, is going to result in increased skin cancer levels, even in parts of the world that are not bathed by the sun year round. Like ours. So we'd all better slip, slap, and slop. The slap is easy: just put on a wide-brimmed hat. The slip and slop are more difficult.

You would think that any shirt would do, but ultraviolet light can penetrate fabric, especially if it is white and the garment fits tight. Loose, dark-colored T-shirts are best. And, for even greater protection, slip on some of the new clothing made from fabrics treated with chemicals that block all ultraviolet light. The benefits of covering up were clearly shown by a team of Israeli researchers who compared skin cancer rates of recent European immigrants to Israeli and Orthodox Jews of similar ancestry. Orthodox Jews dress in such a way that only the skin of their hands and faces is exposed, and the men wear wide-brimmed hats. Not surprisingly, skin cancer rates among this group proved much lower.

Even Miss Coppertone has taken heed of such studies. Remember those ads that featured the cute little girl who'd had some kind of conflict with her dog? I don't know what she did to the animal, but it retaliated by nipping at her rear end. Luckily, its teeth caught the swimsuit, not the child, and the dog tugged the suit down just far enough to reveal a white derriere in sharp contrast to the deep tan elsewhere. Today's Miss Coppertone has slapped on a hat and slipped on a T-shirt. In deference to our current sensitivities, the dog tugs on the T-shirt, not the swimsuit. So we can no longer compare the exposed and the unexposed parts of the anatomy, but I suspect there would not be much contrast. The little girl looks pale everywhere. Thanks to the Coppertone lotion she is carrying — the lotion she has obviously slopped on.

There is some neat chemistry here, but to get a handle on it we first have to understand something about ultraviolet or "black" light. My first exposure to the nuances of this invisible form of radiation happened many years ago, and it came about in an unusual way. At the time laundries marked shirts for sorting with dyes that were invisible under normal light but fluoresced under ultraviolet lamps. These laundries, of course, did not foresee the advent of discos, dance clubs whose operators exploited the bizarre effects that black light could produce. I found out about it the hard way when the laundry numbers on my shirt suddenly started to flash on the dance floor. Haven't been in a disco since. But our concern here is not with what ultraviolet light can do to shirts, but what it can do to our skin. And that all depends upon what kind of ultraviolet light we are talking about.

Just like visible light, ultraviolet light is made up of a range of wavelengths. What does that mean? Think of what happens when visible light passes through a prism. It separates into the colors of the rainbow. These colors are just different types of light waves, each having a characteristic wavelength. Similarly, ultraviolet light is composed of a range of wavelengths. The waves are very short, so short that they are measured in tiny units, called nanometers. In general the wavelength range for sun exposure is from two hundred to four hundred nanometers. The shorter wavelengths are the most dangerous. Those in the 200-to-290 range are referred to as UV-C, but these are essentially filtered out by the ozone layer in the atmosphere. However, as the ozone layer develops those holes that we've heard so much about, UV-C starts to become a problem. UV-B is defined by wavelengths in the 290-to-320 nanometer range. Not only are these the rays that produce sunburn, but they are also the ones that stimulate cells called melanocytes to produce the pigment melanin, which accumulates in the skin to produce a tan. A tan is the body's attempt to reduce further damage by having melanin act as a sunscreen. There is therefore no such thing as a safe tan. If our skin is manufacturing an excess of melanin, then the sun has already done us some damage.

To avoid sustaining this damage we need to cover ourselves with a substance that either reflects ultraviolet light or absorbs it. Our sunscreens used to contain chemicals that absorbed UV-B, the shorter wavelength, and the more energetic rays that we knew caused burns. A slew of compounds, ranging from para-aminobenzoic acid (PABA) to homomenthyl salicylate, performed this job nicely. We could even blend them in different ratios to achieve various levels of protection. And so SPF, for "sun protection factor," entered our vocabulary. Those who applied a lotion with an SPF of eight could expect to stay out in the sun eight times longer than they could with unprotected skin before burning. So we happily slopped on sunscreens that vied with each other to provide greater and greater protection. The numbers soared, as did the price — SPFs of fifteen, thirty, and sixty crowded drugstore shelves.

They did the job as long as we used them properly. That meant selecting a lotion with minimum SPF of fifteen for short exposures and thirty for longer ones (the added benefit of lotions with an SPF of more than thirty remains questionable); we also had to apply the stuff generously enough — roughly two teaspoons to cover the face and arms. If we followed these rules, we didn't burn. But then we found out that those longer UV-A rays that we hadn't worried about because they didn't signal their presence by visibly roasting our skin were nevertheless wreaking silent havoc. They were triggering skin cancers and wrinkles. These also had to be blocked. So oxybenzone, benzophenone, and avobenzone (sometimes called Parsol 1789) entered the scene. Again we thought we were safe. Until we learned that these compounds could actually deteriorate in sunlight and even degrade the UV-B blockers. Chemists came to the rescue, though, by adding octocrylene, which protected the active ingredients from breaking down with sun exposure. Sort of a sunscreen for sunscreens. And then another concern emerged. Just what had happened to all that energy from sunlight that our sunscreens were absorbing? The experts conducted a few experiments, and the results suggested that the molecules dissipated the extra energy by forming some of those notorious "free radicals" that can damage skin. Sensational headlines about sunscreens causing cancer appeared, triggering yet more public confusion. Why not reflect the light instead of absorbing it? We knew that titanium dioxide and zinc oxide could do this effectively, so the chemists incorporated them into the next generation of sunscreens. Still newer versions make use of microcrystalline forms of these compounds, and that means they no longer leave us looking like we've been dipped in white paint.

And what's new under the sun in terms of sunscreens? The reddish secretion on the back of a hippo is a very effective sunscreen that researchers have discovered works well on humans. But I imagine it will be a while before the public accepts hippo sweat as a sunscreen ingredient. Alpha-glucosylrutin (AGR) may be more marketable. This plant derivative has potent antioxidant properties and neutralizes those nasty free radicals that sunlight can generate. Putting antioxidants into sunscreens is nothing new; we've used vitamins C and E for this purpose. In fact, a study at Duke University showed that a vitamin-C laced suntan lotion prevented pigs from being burned by ultraviolet light that would damage unprotected skin. Human-study results are more ambivalent, and many dermatologists argue that the antioxidant vitamins do not penetrate the skin sufficiently to offer any significant protection. A more exciting approach would be to infuse sunscreens with enzymes that can actually repair damage to DNA molecules caused by ultraviolet light.

When we encapsulate these enzymes in little beads of fat (then referred to as photosomes) they penetrate tissue. This formulation proved capable of repairing damaged skin when researchers tested it on human volunteers.

There's something else we can do to protect ourselves from the sun: eat right. Study volunteers who ate twenty-five milligrams of mixed carotenes daily experienced a significant reduction in burn intensity when exposed to ultraviolet light. A diet in which fewer than twenty percent of the calories derive from fat has also been linked with lower rates of skin cancer. Still, the best protection doesn't come from diet or, indeed, from sunscreens. It comes from the shadow rule. If your shadow is shorter than your actual height, look for shade. Or, if you prefer to listen to Noel Coward, remember that only mad dogs and Englishmen go out in the noonday sun. And perhaps a few orphaned baby elephants.

Cures Me, Cures Me Not

The Dominican Republic beach was sprinkled with vendors hawking delicious peeled oranges, raw oysters, and various shlocky souvenirs. But the little stand that caught my attention displayed rows of bottles that seemed to be stuffed with leaves and twigs. My wife and I wandered over to look. "What are these," I inquired? "Medicine," came the reply. "For what?" "Colds, headaches . . . everything." Judging by the appearance of the vendor this was a bit of an exaggeration — the stuff clearly wasn't good for maintaining dental health or growing hair. Nevertheless, I pressed on. "How do you know?" The man explained to me that everyone in his family used it; it kept them healthy and gave them energy (something he seemed to be lacking). "How do you take it?" "Put into hot water and drink."

"How much of it?" "As much as you need." "How often?" "As often as you need." "What's in it?" "The purest herbs!"

By now the man had sensed my growing skepticism, so he brought out the heavy artillery. He motioned me over to the side of his stand and indicated to my wife that she should busy herself with some jewelry he also had on display. This was not a problem. The vendor then picked up a small statue, which appeared to be of a man with a barrel slung over his shoulders. My new friend then placed an arm around my shoulder and with his free hand lifted the barrel on the statue. The little guy's manhood sprang to attention. The message was clear. The stuff in the bottle really was good for everything. I bought it — purely as a souvenir, of course. (I was not allowed to buy the little statue.)

But do you know what? You don't have to travel to the Dominican Republic to be subjected to herbal confusion. You can experience it all right here. Magazines and television ads, multilevel marketers, and numerous Web sites tantalize us with miracle concoctions that have little more scientific backing than my Dominican panacea. It's a shame, because herbal remedies deserve better treatment. Many have a great deal to offer, albeit rarely as much as their vigorous promoters would have us believe. Why are herbal remedies so widely embraced today? It's partly due to a growing disenchantment with modern medicine. Costs are too great, cures are too few. People feel cheated because modern science still has not discovered the secrets of perpetual health or eternal youth. They distrust the pharmaceutical industry and its "synthetic" drugs. By contrast, they see herbal remedies as a safe, time-honored "natural" alternative. But it's not so simple.

Let's take the case of one of the most popular current herbal therapies: echinacea. Way back in 1871 Dr. H.C.F. Meyer of Pawnee City, Nebraska, discovered that the Sioux Indians were using a preparation made from a purple coneflower to treat various ailments, including painful teeth and colds. They would also apply poultices made from the plant's root to wounds, including rattlesnake bites. The good doctor heard opportunity knock, and so he created Meyer's Blood Purifier. Since the substance was supposed to be effective against snakebite, people began referring to it as "snake oil." But since it didn't work very well for this purpose, the term "snake oil" came to represent any heavily hyped but ineffective medicinal product. While extracts of echinacea (the botanical name for coneflower) may not have done much for snakebite, they did appear to have an effect on some other ailments, particularly colds and flu. Indeed, echinacea became the most popular plant sold in North America until sales plummeted in the 1930s, when the first effective antimicrobial preparations, the sulfa drugs, entered the scene. These days, because we have become concerned about the overuse of antibiotics and because natural therapies in general have a growing allure, sales of echinacea are brisk again.

So does echinacea work? One would think that this would be a relatively simple question to answer, but it isn't. The best reply I can come up with is that some echinacea preparations work for some conditions in some people some of the time. I realize that this is not very satisfying, but such is the nature of the herbal beast. First of all there are nine species of echinacea, although manufacturers only use three (E. purpurea, E. angustifolia, and E. pallida) to make supplements. Each of these plants has a different chemical profile. Each contains dozens of compounds; some they have in common, some they don't. Furthermore, their leaves, flowers, stems, and roots have different compositions. An alcohol extract of the root will have a very different chemical makeup from a hexane extract of the stem or from capsules filled with dried, powdered leaves. Before we even think about carrying out tests on the efficacy of echinacea, we must make a decision about what we are going to test. In other words, we require a standardized product. But standardized for which ingredient?

Determining the extent to which dozens of compounds are present in these substances is a practical impossibility. We should test for the presence of the compounds we think are physiologically active. Unfortunately, in the case of echinacea there are many candidates. Polysaccharides, isobutylamides, caffeic acid, cichoric acid, and echinacosides are just some of the possible active substances. Manufacturers who standardize their products do so usually in terms of these compounds, which we collectively refer to as phenolics. But when Consumers Union scientists carried out a survey, they found that the actual range of phenolics in different products was 0.8 to 4.5 percent, even though all the bottles were labeled as containing four percent phenolics. This means that people taking different products could be getting dramatically different doses of active ingredients, even though the products claim to be standardized in the same fashion. Still more disturbing to the Consumers Union team was the fact that different bottles of the same brand sometimes had different amounts of the chemicals. So, based upon this finding, it isn't surprising that some studies show effectiveness and some do not.

There is no doubt, though, that echinacea has physiological activity. Researchers have cured rodents infected with bacteria by injecting them with echinacea solutions. But this is not the same as a human taking an oral preparation. The polysaccharides, which are widely regarded as active molecules, probably do not survive digestion. Test-tube studies also indicate that a number of echinacea preparations can increase the activity of certain white blood cells that have immune function, but human evidence is harder to come by. Conflicting studies abound. Perhaps the most compelling positive study used a product called Echinaforce; patients who took it reported that their cold symptoms improved by twenty-five percent, as compared to the results with a placebo. On the one hand, we have little evidence that echinacea can prevent a cold. On the other hand, we have little evidence that it can do us much harm. Not even to pregnant women, as a study conducted at Toronto's Sick Children's Hospital demonstrated. But people who suffer from autoimmune diseases, such as lupus, multiple sclerosis, and arthritis, could conceivably suffer adverse effects if echinacea does what it is supposed to do — namely, enhance immune function. Finally, some pediatricians claim that they have been able to cut down on prescribed antibiotics for ear infections by substituting echinacea, while others have not found echinacea to be effective. Perhaps they were using different products.

Echinacea is certainly an interesting substance, but there are too many variables in the products it spawns and too many conflicting studies to allow for generalized recommendations. Just consider the following studies, all of which have been reported in the scientific literature. When researchers bathed cells in echinacea extracts and exposed them to influenza and herpes viruses, only a small proportion became infected, compared with untreated cells. Yet sperm exposed to high doses of echinacea extract showed DNA damage and impaired viability. In a German study women with recurrent vaginal yeast infections reduced their recurrence rates significantly by using an antiyeast cream supplemented with oral echinacea. In another German study researchers demonstrated that an intake of four milliliters of echinacea just twice daily reduced the incidence of colds by thirty-six percent; but in an American study of one hundred Seattle-area residents who were taking echinacea these subjects had more respiratory infections than members of a control group on a placebo. A Swedish study showed that people who consumed EchinaGuard (one of the best-tested echinacea preparations) got rid of their colds in six days — on a placebo it took twelve days. But in Germany, colds lasted seven days whether the test subjects took EchinaGuard or not. Confusing, isn't it?

So what are we to make of a product like Dr. Jamieson's Instant Chicken Soup fortified with echinacea? According to the package, each serving contains one gram of echinacea root. How much active ingredient does this translate into? Nobody knows. Has anyone conducted studies to prove that the product can vanquish the common cold? It's doubtful. Still, it couldn't hurt to try some. Maybe I'll do that the next time I have a cold. If it doesn't work, perhaps I'll try Echinaforce. Or plain old homemade chicken soup. In any case, I think I'll pass on the Dominican herbal miracle. Stimulating advertising, but no hard facts.

The Crab Shell Game

Eating crabs is a tough business. First you have to tear the creatures limb from limb for the sake of a few tasty morsels. Then there's the mess you make by dipping those bits of liberated flesh into butter. And what are you left with? A pile of guilt, a heap of mangled, empty crab carcasses, and some interesting chemistry.

The exoskeleton of the crab and other crustaceans is primarily composed of a substance called chitin. This is a giant molecule made up of fundamental units of N-acetyglucosamine strung together like links in a chain. Chitin is abundantly available, thanks to a crab and lobster industry dedicated to making it easier for us to consume their products by removing the meat from the shells. A commercial use for chitin would therefore be most welcome. Actually, chitin itself is quite useless, but we can readily convert it into a substance called chitosan by heating it in an alkaline solution. This process removes the acetyl groupings from the molecule, leaving a long chain made up of glucosamine units. If chitosan is added to a container of water, it will form a kind of gel and sink to the bottom. Since floating impurities are carried down with the chitosan, it can be used to clarify beverages. Some chemists have even attempted to produce a plastic wrap from chitosan, and prototypes have been successful in preventing frozen pizza crusts from turning soggy.

But the real excitement about chitosan began when researchers noted that it could not be digested by rats. What went in came out. In fact, more came out than went in. Chitosan had somehow combined with fat in the rodents' digestive tracts and spirited it out of their bodies. Could this be the dietary miracle that weight-conscious people had been seeking for so long? Could we really have our cake and eat it too? Human studies soon showed that obese patients who consumed large amounts of chitosan did lose weight in the short term, and that was enough for the hucksters to capitalize on. Never mind that nobody had done any long-term studies. Never mind that nobody knew whether chitosan had untoward effects, such as interfering with the absorption of fat-soluble nutrients. TV infomercials started to appear, touting chitosan as a wonder substance that would open the door to penalty-free indulgence in cakes, cookies, hamburgers, and french fries.

The hucksters provided a demo. Viewers watched in amazement as a demonstrator sprinkled chitosan onto a pool of oil that was floating in a glass of water. The oil thickened and sank to the bottom. You didn't have to stretch your imagination too far to envision the same kind of magic taking place in your gut.

Take some chitosan, stuff yourself with cake, and the fat will vanish from your body. Hey — not so fast!

While that demonstration is an attention-grabber, it isn't realistic. You can't stuff yourself full of fat and expect chitosan to save you from dietary damnation. There's a limit to the amount of chitosan that the body can handle comfortably. One can easily handle a few grams a day, and the stuff will prevent the absorption of some ten to twenty grams of fat, but that would hardly make a dimple in the fat content of the cornucopia of rich foods the hucksters insist we can consume without guilt. Clearly, chitosan is no miracle. There is no fat-free lunch. But chitosan is not a total dietary dud, either.

Dr. Arnold Fox may have the right idea about how to use chitosan. In his somewhat sensationally titled Fat Blocker Diet he prescribes a program of exercise coupled with a low-fat diet that allows for the occasional splurge. This is where the chitosan comes in. According to Fox, most low-calorie diets fail because they foster cravings for fatty foods. So have the odd sundae, he says, but take the chitosan first. Fox reels off plenty of anecdotal evidence gleaned from his years of medical practice to buttress his argument. His approach may not be scientifically well documented, but it is still far more reasonable than that of the chitosan hucksters who offer the simple-minded suggestion that we gobble four to eight capsules of the stuff daily while gorging on high-fat foods. The latter approach may sound attractive to some, but a well-designed, placebo-controlled study of thirty subjects published in *The European Journal of Clinical Nutrition* has shown that it just doesn't work.

While chitosan may not be a major player in the battle of the bulge, it may be of some use in the battle against osteoarthritis. This painful condition arises when the rubbery tissue found at the ends of bones, known as cartilage, begins to wear away. As we move and exert pressure on our joints, cartilage provides a cushioning effect by absorbing and releasing some of the synovial fluid that normally surrounds the joints. When the cartilage degrades, bone can rub painfully on bone. Standard treatment usually involves nonsteroidal anti-inflammatory drugs, which may check the pain, but they do not impede the progress of the disease. They can also cause gastric problems.

There may be a better, gentler approach: chitosan — or, more specifically, glucosamine, the chemical degradation product of chitosan. Scientists became interested in this substance when they discovered that it was one of the building blocks of proteoglycans and glycosaminoglycans, fundamental components of cartilage. Glucosamine was also a component of hyaluronic acid, the substance that makes synovial fluid thick and elastic. The theory, which appealed to European researchers years ago, was that a supplemental intake of glucosamine would enhance the formation of hyaluronic acid and also provide cells in cartilage, known as chondrocytes, with the raw material they needed to repair the cartilage.

This idea did not sit well with American scientists, who insisted that glucosamine taken orally would degrade to glucose before it ever reached a joint. The Europeans didn't pay much attention and tried it anyway. It worked. Glucosamine was not a wonder drug, but in thirteen major studies it proved to be better than a placebo. It worked as well as, or in some cases, better than, the nonsteroidal anti-inflammatories. It even worked on dogs and horses. Recently, Chinese researchers conducted a study in which subjects were given fifteen hundred milligrams of glucosamine daily for four weeks; it relieved pain better than ibuprofen and caused far fewer adverse reactions. The consensus of the studies carried out so far is that some osteoarthritis sufferers will certainly be helped by glucosamine. Unfortunately, however, the overall results, while meaningful, are not very impressive. In one of the best studies so far Dr. Joseph Houpt of the University of Toronto recorded that forty-nine percent of the subjects claimed to feel better after taking glucosamine. But Houpt also noted that forty percent of those taking the placebo also felt better. A somewhat more optimistic picture was painted by a study published in January 2001 in *The Lancet*, one of the premier medical journals in the world. Researchers found that not only did glucosamine alleviate the symptoms of osteoarthritis, but it actually reduced the rate at which cartilage deteriorates. There are no major side effects associated with the use of glucosamine, although some patients have complained about nausea. There is also some concern that it might alter insulin requirements in diabetics.

Sometimes glucosamine is paired with chondroitin sulfate for the treatment of osteoarthritis. Chondroitin is also an integral part of cartilage, mostly responsible for drawing fluid into the tissue to give it elasticity. It also inhibits the enzymes that break down worn cartilage. But the idea that oral chondroitin, which is isolated from cow cartilage, should be an effective treatment is a highly speculative one. Chondroitin is a very large molecule that will, in all likelihood, be degraded before it can reach its target. But let's not discard a treatment because it isn't theoretically plausible. Indeed, some recent studies have shown that glucosamine with chondroitin works better than glucosamine alone. Still, glucosamine, or the combo, does not work for everyone. If a patient who takes it doesn't experience improvement after three months, it probably won't happen at all. And there is one more problem. Dietary supplements like glucosamine are very loosely regulated, and there is no guarantee that the pills you buy actually contain what they say they contain. But maybe I'm just being crabby.

Wondering about Wort and Grapefruit

The plant with the pretty yellow flowers was christened St. John's wort because it begins to bloom around June 24, the purported birthday of St. John the Baptist. Until a couple of years ago, *Hypericum perforatum* grew on our roadsides in virtual obscurity, but today preparations made from the plant jostle for our attention on the shelves of pharmacies and health food stores. "Relieves insomnia, neuralgia and nervous tension," one label boldly declares, while another sedately promises to make you "feel your best." Sounds intriguing.

The use of St. John's wort as a medicinal substance dates back to the Middle Ages ("wort" is a medieval English word for "plant"), when the substance was recommended as treatment for anxiety and bad dreams. While we cannot assume that a plant has true medicinal value just because we've been using it for a long time, its long tradition should spark scientific investigation. After all, many of our current drugs evolved from research stimulated by the folkloric use of plants. Digitalis, a preparation made from the foxglove plant, was the first effective treatment for congestive heart disease. It was introduced in the eighteenth century by William Withering, an English physician who recognized foxglove as the key ingredient in an old wive's remedy.

Today foxglove is still the source of digitoxin, a widely used drug that increases the force of the heart's contraction. To produce it chemists separate and purify one compound from the plant's roughly thirty cardiac glycosides, each of which has different activity. Manufacturers of herbal supplements do not note such distinctions; they merely try to justify their products by pointing out that many of our prescription pharmaceuticals derive from plant sources.

St. John's wort, like all plants, produces an array of compounds, which can account for its antianxiety and antidepressant effects. These effects are more than anecdotal; they are supported by a number of studies. Most of these studies have been carried out in Germany, where doctors frequently prescribe standardized versions of the plant as an antidepressant. They actually write more prescriptions for Johanniskraut, as the plant is known in that neck of the woods, than for Prozac. In light of a meta-analysis — a study of studies — published in 1996 in the British medical journal *The Lancet*, the choice of those German doctors seems justified.

The meta-analysis covered twenty-three randomized trials involving 1,757 patients, and its overall conclusion was that St. John's wort is significantly better than placebo. Although the plant contains numerous compounds that are candidates for biological activity, scientists now think that hypericin is the major therapeutic agent. That is why manufacturers have standardized preparations to contain 0.3 percent of this compound. The usual recommended dose is three three-hundred-milligram standardized capsules three times a day; it usually takes several weeks for the effect to manifest itself. St. John's wort should not be taken along with other antidepressants, especially so-called serotonin re-uptake inhibitors like Prozac. Because the two have a similar action, patients who combine them may experience confusion, shivering, and muscle spasms — all signs of serotonin syndrome.

St. John's wort raises other concerns, as well. First of all, no one should attempt to diagnose and treat their own depression. Then there is the fact that most St. John's wort studies lasted only a few weeks, so long-term effects remain undetermined. In the short term some people may experience side effects such as upset stomach, skin rash, or tiredness, although these are less common with St. John's wort than they are with prescription antidepressants. Still other people may find that their skin is more sensitive to sunlight. After taking all of this into account, a physician may nevertheless conclude that St. John's wort is an appropriate initial treatment for mild depression, anxiety, or sleep problems. An herbal therapy that really works. But — and there is a but — consider the following account.

Doctors at the Zurich hospital were stymied. Their sixty-one-year-old heart transplant patient had been doing very well for a year, but he was now complaining of constant fatigue. Blood tests presented only one clue. His plasma level of cyclosporin was lower than it should have been. Like all transplant recipients, this patient had been placed on a regimen of drugs, including cyclosporin, that suppress the immune system — this should prevent rejection of the transplanted organ. The low cyclosporin level prompted the doctors to order a biopsy of the heart muscle, which, unfortunately, showed that the man's body was indeed rejecting the new heart. But why had his blood levels of the drug, which had been normal for nearly a year, dropped so suddenly? What had changed? Was the patient doing anything different? As it turned out, he was.

A few weeks earlier he had diagnosed himself with mild depression, and he'd known exactly what to do about it. He'd seen stories about the potential benefits of the herbal remedy St. John's wort in all kinds of tabloids and magazines, and here was his chance to give it a try. It never crossed his mind to check with his doctors first, since St. John's wort was a "natural" remedy, and therefore (supposedly) completely safe. But the effects of drugs on our bodies can be very complex and sometimes mysterious. While the mechanism by which St. John's wort carries out its antidepressant effect is not clear, researchers have been able to identify an unwelcome side effect. Some component in St. John's wort stimulates the production of an enzyme with the cryptic name of CYP3A, one of a number of enzymes collectively called the cytochrome P-450s. What are enzymes? They're specialized protein molecules that serve as biological catalysts, speeding up chemical reactions that take place in the body all the time. The reactions pertinent to this case involved the breakdown of substances the body perceives as foreign. Many drugs fall into this category. Indeed, pharmaceutical researchers contend with this fact all the time, and they have to design dosages that take into account that some of a drug — cyclosporin, for example — will be lost to cytochrome activity. But what happens when cytochrome activity increases in an unanticipated fashion? Complications arise.

Our heart transplant patient took a standard dose of St. John's wort for three weeks. This resulted in an increase in CYP3A, which, in turn, led to a drop in his blood levels of cyclosporin. His immune system, previously held in check by cyclosporin, now took aim at the foreign heart and began rejection proceedings. Luckily, his doctors caught the problem in time, stopped the St. John's wort, and administered powerful antirejection drugs intravenously for ten days. The man survived, but I think that his days of self-diagnosis and self-prescribing are over.

If St. John's wort can alter levels of cyclosporin in the blood, might it not also interfere with the action of other medications? Recent research indicates that it can. Not surprisingly, the affected drugs are those that, like cyclosporin, are also metabolized by cytochrome enzymes. Protease inhibitors, used in the treatment of HIV infections, are a prime example. Because of the popularity of St. John's wort as an antidepressant and the incidence of depression in patients diagnosed with HIV infections, researchers at the U.S. National Institutes of Health decided to investigate the consequences of using the herbal remedy and the protease inhibitor indinavir concurrently. Doctors prescribe indinavir to prevent the HIV virus from reproducing, but the dosage is critical. If blood levels are not optimal, then treatment becomes ineffective, and the chances of the virus developing resistance to the drug increase. So the researchers recruited healthy volunteers to take indinavir and St. John's wort at the same time. The results were dramatic. Concentrations of indinavir in the blood fell by roughly fifty percent when subjects took the drug along with St. John's wort.

The National Institutes of Health study highlights the importance of disclosing all of the medications you are taking, whether they be "natural" or not, to your prescribing physician. You may be compromising the effectiveness of a number of other common medications if you are taking them concurrently with St. John's wort. These include coumadin, a common blood thinner; digoxin, used in the treatment of congestive heart disease; and antidepressants of the serotonin re-uptake inhibitor class. Women taking oral contraceptives could be in for a surprise, because St. John's wort reduces blood levels of a contraceptive's active ingredients, making an unintended pregnancy possible.

All of this is just what we needed — something else to worry about. It's enough to make your blood pressure rise. But of course there are pills for that, too. How do you take them? If you want to play it safe, don't wash them down with grapefruit juice. Here we go with those pesky cytochrome enzymes again.

Felodipine is an effective blood-pressure-lowering drug, but before seeking approval to put it on the market researchers wanted to check out any possible adverse effects for patients taking it with alcohol. At the University of Western Ontario they designed a double-blind study using grapefruit juice to mask the taste of the alcohol. To the researchers' great surprise, the alcohol had no effect, but the grapefruit juice increased the drug's potency significantly. It turned out that some component of the grapefruit juice inhibits the enzyme that normally degrades felodipine, a cytochrome P-450, an effect exactly the opposite of the one attributed to cyclosporin.

This finding was not without consequence. Essentially, it meant that people who took the medication with grapefruit juice might see their blood pressure drop below the desired level. But it also meant something else. If researchers were able to isolate the ingredient in grapefruit juice responsible for blocking the enzyme, then they could, in theory, incorporate it into the medication. Doctors could then prescribe lower doses, reducing side effects but still achieving the same degree of effectiveness.

While this is a tantalizing possibility, the complexity of the chemistry involved probably precludes it from being put into practice in the near future. Compounds called flavonoids, present in grapefruit juice, appear to be responsible for the effect, but indirectly. The grapefruit factor is extremely variable, with people experiencing different results even after drinking the same amount of juice. Researchers have found that the actual enzyme-blocking chemicals are not present in the juice but form upon metabolism of the flavonoids. Their exact nature has not been identified, but what we do know is that they can potentiate drugs other than felodipine. Some cholesterol-lowering medications, some antihistamines, some hormone supplements, and even antirejection medications can be affected.

So what lessons do we draw from this account? Don't assume that herbal supplements are safe under all conditions. Always tell your physician everything you're taking. If you've been taking medications with grapefruit juice and they have been working well, don't change anything. But don't start taking any new medications with grapefruit juice. Orange juice is fine. What about drinking grapefruit juice at other times? Studies show that the active compounds stay in the blood a long time, so if you are taking medications, don't go overboard with the grapefruit juice. Of course if our heart transplant patient had done exactly that, then we may never have found out about the reaction between St. John's wort and other drugs. The two effects may have canceled each other out. And, finally, if you're on the birth control pill and are feeling depressed for some reason and decide to have a go at St. John's wort, it wouldn't be a bad idea to take it with grapefruit juice. That just might neutralize the birth-control-neutralizing effect of the wort. But I wouldn't count on it.

Nothing to Rave About

First there was nausea. Next came the shakes. Then the teenager dropped to the floor like a sack of potatoes. When he came to, he found himself surrounded by medical personnel, all of them mystified as to why a healthy young athlete would faint. They probed him with questions until the boy sheepishly revealed that he had bought some stuff from "Doc's Gym" to make him feel good and bulk up. The Doc of Doc's Gym — located in Montgomery, Alabama — turned out to be a medical doctor named Speer whose license had been revoked for abusing his prescription-writing privileges. There was a small matter of prescribing narcotics when they were not medically indicated. Now the inventive Doc Speer was at it again, but this time he was trying his hand at chemistry, synthesizing a substance called gamma hydroxybutyrate (GHB) in his home.

This was the "stuff" that the teenager had purchased at the gym. Although it's now an illegal substance, GHB was once sold in health food stores as a "natural" dietary supplement to induce sleep and build muscle. Supposedly, the longer periods of sleep it produced led to the release of human growth hormone in the body, a hormone linked with increased muscle mass. But there were so many reports of nausea, shaking, and even coma among GHB users that the U.S. Food and Drug Administration stepped in and outlawed the drug.

This, however, did not put an end to the abuse of GHB. The compound is just too easy to make. Anyone with a basic knowledge of chemistry can cook it up. All you need is a commercially available solvent called gamma butyrolactone (GBL) and some sodium hydroxide, available as a drain cleaner. The Internet offers up plenty of recipes that even a neophyte can follow. Doc busied himself brewing large batches of GHB and selling the stuff in his gym. And he wasn't alone. Gamma hydroxybutyrate is easy to find on the underground market. It's part of the rave scene. Kids take it to reduce anxiety and induce feelings of euphoria. Unfortunately, some have paid a penalty — the ultimate one.

Among them was a seventeen-year-old Houston honor student who had never taken drugs before. She went with friends one night to a local club. Around midnight she complained of a headache and went home. When she didn't come down for breakfast, her mother went to check on her. The girl seemed to be in a deep sleep, and could not be roused. Her breathing was shallow and labored. By the time the ambulance arrived she had gone into cardiac arrest, and, in spite of heroic efforts to save her, she passed away. No drugs were found in her blood, but when the police heard that GHB was circulating in the Houston area, they asked the pathologist to carry out a specific blood test for GHB, a test that is not routinely performed. Sure enough, it came back positive. The girl's death was labeled a potential homicide.

Why homicide? Because GHB has a reputation for being a date-rape drug. The aggressor can easily mix it into a beverage, and, if the dose is right, it will induce a state of stupor, leaving the victim powerless to resist. The effect wears off in a few hours, and the victim's memory of the event is essentially wiped out. Ideal for the criminal. Yet if the dose is too high, the results can be lethal. This is a rarity, but various adverse reactions are relatively common. One of the problems is the purity of the drug; those who manufacture it use sodium hydroxide in its synthesis, and they must carefully remove any excess from the final product because sodium hydroxide is very corrosive.

Just ask the twenty-nine-year-old man who found out about this the hard way. He ended up in a hospital emergency room in an unresponsive state half an hour after ingesting GHB contaminated with sodium hydroxide. He had vomited profusely, and his lungs were damaged when he aspirated some of the sodium hydroxide from his stomach into his lungs while vomiting. His doctors had him in intensive care and were considering him for a lung transplant when his condition finally improved. He eventually recovered. The guy probably doesn't ever want to hear of GHB again.

But lots of people and some scientists do want to hear more about GHB, because the drug has another side. A potentially useful one. Indeed, gamma hydroxybutyrate was created as a possible antianxiety agent, modeled after a naturally occurring substance in the central nervous system called gamma amino butanoic acid (GABA), which has calming properties. However, GABA was not suitable as a medication because it couldn't cross the membrane that protects the brain from intruders, the so-called blood-brain barrier. But GHB could. The man who recognized this was no fly-by-night underground chemist — he was the noted French researcher Henri Laborit, who had developed chlorpromazine, the first effective drug against schizophrenia. Later GHB itself was found to occur in the brain, probably acting as a neurotransmitter that relayed information from one nerve cell to the next. Laborit became a champion for GHB and advocated its general use as a feel-good substance, a sleep inducer, an antidepressant, and an anxiety reliever. He, himself, took GHB regularly throughout his life, calling it his "aurum potabile," or "drinkable gold."

In Europe GHB is available as a drug and has a good record when taken in prescribed doses. Proponents claim that it stimulates sociability, dissolves paranoia, and increases feelings of sexual attraction. It enhances the zest for life, even in people who have previously shown suicidal tendencies. How is it, then, that a drug Europeans accept and use with virtually no side effects is anathema to North Americans? The answer is social context. Drugs are not good or bad by nature; their attributes and pitfalls depend on how we use them. For whatever reason, no North American pharmaceutical company has chosen to go through the regular evaluatory process required to bring this new drug to market. So GHB has fallen into the hands of the underground chemists. We have subjected it to no regulations, no tests for purity, and no scientifically determined dosage recommendations. And those underground chemists are pretty ingenious. When the authorities began clamping down on GHB, the illicit chemists discovered that the easy-to-purchase industrial solvent called gamma butyrolactone (GBL) was converted in the body to GHB, so they made the switch. They started selling GBL under names like RenewTrient and Revivarant. When the government issued warnings about GBL, they switched to selling the solvent 1,4-butanediol (BD), which, they learned, is converted to GBL in the body. But BD, too, can have nasty effects. A Los Angeles chiropractor made over a hundred rave participants sick when he supplied them with BD-containing Cherry fx Bombs from his private lab. The man was manipulating more than people's backs: he was manipulating chemicals and lives.

It's unfortunate that GHB is mired in so much controversy, because it shows promise as a treatment for narcolepsy, leg cramps, anxiety, alcohol withdrawal, and sleep disorders. Some pharmaceutical companies are looking into it, but even if their research pans out and they are able to transform GHB into a prescription drug, the underground chemists will not be vindicated. They aren't working to improve health; their goal is to make money selling illegal substances. The uncontrolled use of GHB, therefore, will always present a risk. After all, it offers a "natural high" and tantalizes with pseudonyms like "Easy-Lay." The authorities remain vigilant, and that's why Doc Speer of Doc's Gym is modeling striped clothing.

Flushing out the System

I think we should urinate more often. At least that's what recent research into the incidence of bladder cancer would imply. A study of some fifty thousand men, reported in *The New England Journal of Medicine* in 1999, showed that those who consumed more fluids, and presumably urinated more often, had a lower incidence of the disease.

Bladder cancer is North America's fourth leading type of cancer, affecting men about three times as often as women. It became one of the first cancers related to a specific occupation, when, in 1954, researchers discovered that men working in the dyestuff industry were more likely to contract the disease. Later, researchers also connected cigarette smoking with bladder cancer. Both these observations suggested that the problem was triggered by a chemical agent that came into contact with the bladder. Scientists soon pointed the finger at compounds called arylamines, which are present both in dye preparations and in tobacco smoke. Furthermore, arylamines induced cancer in test animals. But these are not the only carcinogens involved. Painters, truck drivers, dry cleaners, and drill-press operators also manifest a higher incidence of bladder cancer, presumably due to chemical exposure. So it came as no surprise when, in the 1980s, Israeli scientists suggested that urban men, who urinated less, had higher rates of bladder cancer than rural men, who drank more and urinated more frequently. Urban urine was more concentrated, the researchers claimed, and prolonged storage promoted contact of carcinogens with the bladder. This hypothesis now appears to be substantiated by the study reported in *The New England Journal of Medicine.*

But, given that not all bladder cancer victims smoke, work with dyes, or operate drill presses, where could the carcinogens be coming from? Chlorine-treated drinking water is a possibility. Epidemiological surveys have revealed that people who drink surface water treated with chlorine are more likely to develop bladder cancer. How much more likely? Estimates are that in North America we can link roughly 4,500 cases of bladder cancer annually to chlorinated water. To put this into more understandable terms, about ten out of every thousand men who do not drink chlorinated water will develop bladder cancer if they live to be seventy, while thirteen will do so if they drink chlorinated water for thirty-five years. This, of course, assumes that those who don't drink chlorinated water are consuming water that has been purified by some other technique, an option not available to most people.

But to retain the proper perspective here, let's remember that chlorination is probably the most important public health measure in the history of the world. Chlorine was first used to disinfect water at Maidstone, England, in 1897, during an outbreak of typhoid fever. People had known about the pale green gas since 1774, when the German scientist Scheele generated it by treating salt (sodium chloride) with sulfuric acid and manganese dioxide. He noted that the gas dissolved in water and that it had an odor "most oppressive to the lungs." Scheele also observed that a chlorine solution bleached flowers and leaves, and this would turn out to be an extremely useful observation. The cotton we spin is gray, and we have to bleach it before dyeing. The standard method for doing this before the advent of chlorine bleach was called "grassing": workers would moisten fabrics with sour milk and unroll them across large grassy areas to be bleached by the sun. When chlorine bleach was introduced, huge expanses of land were freed for agriculture, because a few properly constructed vats took the place of thousands of acres of drying land.

The workers who did the bleaching had to protect themselves from the acrid chlorine vapors, and they discovered that breathing through handkerchiefs dampened with alkali worked well. Noted chemist Count C.L. Berthollet discovered the reason. Calcium hydroxide, better known as lime, a classic alkali, reacted with chlorine to form calcium hypochlorite — a nonvolatile solid. Berthollet then noted that a solution of this compound made a great bleaching agent, just like chlorine. Soon manufacturers began marketing the solution as Javelle water, and the rest, as they say, is history. We can also employ such bleach solutions as water purifiers: about eight drops of common bleach per liter of water will do the job in an emergency.

Countless lives have been saved since chlorination was introduced in 1897. That's because chlorine is an extremely reactive substance, capable of destroying microorganisms as well as stains. But this reactivity is also responsible for the putative bladder cancer connection. Surface waters, such as rivers, contain a variety of dissolved organic compounds, originating from the breakdown of algae, sewage, and other animal and vegetable matter. Chlorine reacts with these to form organochlorides, some of which are carcinogens. The most worrisome ones have been christened "trihalomethanes" (THMs). Of these, chloroform is the most prevalent and the most widely studied. Yes, this is the same chloroform that was used as an anesthetic until relatively recently. I remember having my tonsils out under chloroform anesthesia in 1953. Luckily, that was before it was a carcinogen.

Today chloroform and the other THMs are recognized as carcinogens, and their presence in drinking water is closely monitored. Different countries have established different limits for THMs, but a good general guideline is that we should maintain chloroform below one hundred micrograms per liter. Concern about THMs is much less pronounced in places where chlorination is not the main form of water treatment. The city of Montreal, for example, uses ozone, adding only a small amount of chlorine to protect the water through the distribution system.

So we face a curious conundrum. We should be drinking more to flush carcinogens from our bladders (as well as for other health reasons), but the stuff most of us drink — namely, tap water — may itself contain some of those very carcinogens. What do we do? Bottled waters are not chlorinated and therefore do not contain any THMs. We can also remove these compounds with home water-filtration systems, usually with an efficiency rate of well over ninety percent. Just remember, though, that you have to change the filters in these systems regularly, or you'll end up drinking water that's of poorer quality than it is in its unfiltered form.

These days researchers agree that we should aim to drink roughly eight glasses of water daily. This will increase our urine output, benefiting the kidneys and the bladder. For the few men who suffer from bashful bladder syndrome (the fear of urinating in public), this will pose a problem, but leave it to modern science to come up with an answer. A mathematical one. Bashful bladders, scientists now tell us, can be spurred into action at public urinals if their owners concentrate on doing some simple arithmetic. For them I offer this problem: a chloroform concentration of one hundred micrograms per liter corresponds to how many parts per billion? There, that should do it. Just watch your shoes.

Jumpin' Jimson Weed

The emergency-room physicians were puzzled when the young man complained of blurred vision and a dilated pupil in one eye. But when the patient revealed that he may have gotten something into his eye while weeding with a mechanical trimmer, the doctors had something to work with. They dragged out a book on common weeds and asked the patient to identify any that he had seen in his garden. When he pointed to *Datura stramonium*, the mystery was solved. Jimson weed had claimed another victim. And therein lies a story. Several, in fact, but let's start at the beginning.

The first British settlement in North America was established in 1607 and named Jamestown, after King James I. The little tobacco-growing colony was located on the East Coast, in the region that would eventually become the state of Virginia. One of the constant debates among the Jamestown settlers was whether they should expand the colony. Tobacco growing was a very profitable enterprise, especially since labor was so cheap — the Jamestown settlers had brought the first African slaves to North America. But the settlement was surrounded by unfriendly Indians who would, of course, resist the expansion.

Among the expansion's greatest advocates was Nathaniel Bacon, a member of the governor's council. In 1676 he took the law into his own hands and organized an expedition against the Indians, but Governor William Berkley, fearing a large-scale war, denounced Bacon's actions and sent soldiers to quell what would later be dubbed the Bacon Rebellion. Berkley's soldiers prepared to move against the insurgents. They camped in a field and cooked up a stew, which they flavored with a plant they found growing nearby. A most remarkable scenario began to unfold minutes after they had eaten their meal. All thoughts of battle disappeared. The soldiers began to run around laughing and yelling at each other, their speech slurred. This delirium continued for eleven days. The governor's army had been defeated not by Bacon's men, but by a lowly weed, known to this day as Jamestown weed — or, in its corrupted form, Jimson weed.

Jimson weed contains the naturally occurring compounds atropine and scopolamine, which can interfere with the activity of the nervous system by blocking the action of a key chemical known as acetylcholine. This interference can produce hallucinations. Ingestion of the plant can also cause dilation of the pupils, blurred vision, rapid heart beat, reduction of salivation, as well as sedation; and all of these effects are potentially useful in the practice of medicine. Consequently, by the early 1800s people the world over were buying Jimson weed from their local apothecaries. This included the citizens of Mecca.

According to historical accounts, Mecca's criminal element would entice pilgrims to share a meal with them, a meal that had been prepared with Jimson weed roots and leaves. When delirium set in, the thieves would pounce upon their victims and rob them. The victims would wake up in the hospital with no memory of how they had gotten there — all they knew was that their valuables were missing. They were lucky. A lot luckier than a young lady living in Bogotá, Colombia, who, according to a newspaper account, arrived home on a Monday morning with no recollection of the weekend. Scrawled on her back was the devastating message, "Welcome to the world of AIDS!" Her doctors ordered a medical workup, which revealed that she had been drugged with scopolamine and raped. Fortunately, even though she bore traces of semen from seven different men, she had not been infected with AIDS.

Although we are all horrified by such stories, the effect doesn't last. After all, the events in Mecca took place a long time ago, and Bogotá is a faraway place where, we imagine, drug lords rule and anything can happen. Surely we don't have to worry about anticholinergic poisoning in North America? In fact, we do.

Not long ago the management of a Sheraton hotel in New York called in the police to handle a disturbance. They arrived to find a man staggering around with his arms flailing, shirt unbuttoned, eyes wild, pupils dilated — apparently hallucinating. He could not remember where he had been or what had happened to him. His wallet was missing. In the Bellevue Hospital emergency room, doctors diagnosed anticholinergic poisoning. Something was interfering with acetylcholine activity in the man's nerve cells.

Doing some quick detective work, they identified the man as a jewelry salesman who had been discussing a transaction over a glass of wine with some prospective clients. A chemical analysis of the residue in his wine glass revealed the presence of scopolamine. The poor man had been drugged and his money and jewelry stolen. The scopolamine probably had been isolated from a motion-sickness remedy or from eye drops used by opthamologists. Before they could apprehend the perpetrator, New York police picked up a number of other wild-eyed, stumbling anticholinergic robbery victims.

But, still, this was New York. Sin City. Drug Metropolis. A localized problem. Right? No. Reports of anticholinergic poisoning are on the increase across the United States and Canada.

Most cases don't involve criminal activity: the poisonings can be attributed to the allure of the drug high. Teenagers have discovered that ingesting Jimson weed seeds or drinking tea brewed from the plant can induce a state of euphoria accompanied by hallucinations. A few have also discovered another possible effect of this practice.

A most disturbing report published in *The Journal of the American Medical Association* records the case of four boys who went out into the Texas desert to search for Jimson weed. They just wanted to have a good time. When they failed to return, the authorities dispatched a search party. They found two of the boys sitting at their campsite, hallucinating, oblivious to their friends lying at their feet, overdosed on atropine. This drug is well named. Atropos was the Greek goddess who determined how long a person would live. When a child was born, her two sisters would spin the thread of the infant's life, and Atropos would cut it. For two of the boys in the desert, she cut the thread very short.

One of the problems we face in dealing with *Datura stramonium* abuse is that the weed can be found most anywhere. Some people cultivate it as an ornamental plant. The trumpet-shaped flowers of the Angels' Trumpet — like Jimson weed, a member of the solanaceae family — are prized by gardeners. When it became apparent to city councillors in the Florida community of Maitland that the plants were also prized by teenagers, they took the unprecedented step of issuing a ban on planting it. Horticulturists protested, but perhaps the ban is not such a bad idea. A Canadian couple who had a close call because of Angels' Trumpet would probably agree.

They were at home and had just finished a meal of hamburgers when the man suddenly collapsed. His wife called for an ambulance, but soon after it arrived she, too, lapsed into unconsciousness. Medical personnel suspected carbon monoxide poisoning, but they quickly ruled that out. Both the man and the woman regained consciousness within twenty-four hours, and they were soon able to solve the mystery. The woman, wanting to season the hamburger meat, had reached for a spice container. After adding a little of what it contained to the meat, she realized that she had picked up the wrong bottle — a bottle that held the Angels' Trumpet seeds that she was going to plant next year. She had removed most of the seeds from the mixture, but she'd apparently left several behind, enough so that the couple came dangerously close to hearing real angels' trumpets.

Kissing Toads

Picture a beehive. Then picture a bunch of ugly, warty toads climbing on each other's backs to form a toad pyramid. Now imagine the top toad flicking out his long sticky tongue to catch bees returning to the hive. Or try to visualize one of these ugly creatures attempting to mate with a goldfish and drowning it in the process, or, even more bizarre, paying undue attention to a squashed, obviously dead female toad in the middle of a highway. Sounds like something out of a distasteful cartoon, right? Actually, this is real life in Queensland, Australia. What's going on here? Some apparently ingenious science gone wrong.

In the 1930s the sugar cane industry was just beginning to prosper in Australia. One problem, however, began to plague the sugar cane growers: the grayback beetle. Or, more specifically, the grubs hatching from the beetle eggs that the insects had laid in the cane fields. The grubs' favorite food was sugar cane root. As the beetle infestation spread, the crop withered. Farmers tried to rid their fields of the pest by treating the

ground with carbon disulfide, a fumigant. But the beetles survived and the farmers got sick. Could there be some safe biological control, they wondered?

The ideal solution would be a natural beetle predator. Well, one did exist. Word had come from Hawaii that a species of toad, *bufo marinus*, had already earned the name "cane toad" for its ability to protect sugar cane by dining on grayback beetles. So, in 1935, Australia imported 102 cane toads from Hawaii to drive the pesky beetles out of town. The Australians made a special pond filled with beautiful water lilies to encourage the toads' romantic behavior. The little creatures needed no encouragement, however, and soon the pond was alive with cane-toad tadpoles. When the toads matured, the cane growers deposited them in the cane fields, fully expecting the beetles to succumb to the toads' voracious appetite.

But the beetles could fly and the toads could not. And instead of chasing flying beetles, the toads preferred to sit under streetlights and dine on whatever insects, fried by the light, dropped to the ground. Toad or no toad, the beetles multiplied. So did the toads. Whenever the male toads were not eating — and they'd devour anything that would fit into their mouths, be it a ping pong ball, a bottle cap, or a lit cigarette — they were out searching for females. If none were around, footballs or even human feet might receive their amorous attentions. Soon Queensland was overrun with cane toads.

Within a short time there were reports of chickens becoming sick from drinking toad-infested water and dogs being poisoned after biting into the revolting creatures. By the 1940s the Australians had managed to control the grayback beetle with the pesticide lindane, but the toads had become the scourge of the Outback. What could be done about them? Well, some Australians said that if you can't beat 'em, lick 'em. Or smoke 'em. Rumor had it that the toads produced a hallucinogen that could brighten a humdrum existence.

Some inventive teenagers not only licked the creatures, but they also boiled them in water and gulped down the "toad slime" they exuded. Some dried the toad skins, smoked them, and then waited for the excretions to produce their mind-bending effects. A cheap high, a few exciting hallucinations — that's what the kids were after. What they got were some side effects they hadn't bargained for: delirium, high blood pressure, rapid heart beat, and even seizures. One boy actually suffered a heart attack after downing twenty-five cane-toad guppies on a dare.

Why should toads cause such reactions in humans? Many species of toads have evolved a spectacular defense mechanism that prevents them from becoming a predator's next meal. When frightened, the toads activate glands behind their eyes and secrete a mixture of toxins that will convince an attacker to look elsewhere for its next meal. Scientists have identified as many as fifteen different compounds in toad venom, each capable of causing a ghastly array of symptoms. Bufotalin is a heart stimulant, and, in the form of dried toad skins, was used as a medicine before we learned about the foxglove extract digitalis. Another venom component, bufotenin, is responsible for the effects on the mind.

But how, exactly, does a toad secretion effect the human mind? Bufotenin has a very close chemical similarity to serotonin, a substance used by the nervous system to transmit information from one nerve cell to another. Bufotenin overwhelms serotonin-sensitive cells and triggers effects ranging from hallucinations to seizures. Two Toronto men learned about this the hard way. They ended up in hospital after licking a cane toad they had purchased in a pet shop specializing in exotic animals. And a five-year-old Arizona boy did have a brush with death after he put a Colorado River toad into his mouth. (Just why he did this can only be explained by other five-year-old boys.) In any case, this species, *bufo alvarius*, is the most toxic toad in North America. The youngster developed seizures that had to be controlled with medication.

Bufotenin can literally have a mind-blowing effect. The compound, along with a close chemical relative called dimethyltryptamine, is central to a bizarre procedure long practiced by South American Indians. They prepare a type of snuff from the ground-up seeds of the *Piptadena peregrina* plant and blow it into the nose of a willing recipient through a long tube. Bufotenine and dimethyltryptamine, both of which are found in the seeds, then produce a mind-altering effect. Tragically, in some cases the person blowing on the pipe was too forceful, driving the snuff into the recipient's brain, causing bleeding and death.

What will eventually happen to the cane toads? Some birds have learned to turn the toads over on their backs and eat their tongues. Entrepreneurs plan to manufacture toadskin wallets, belts, and jackets. Souvenir sellers hawk stuffed cane toads. And the Chinese have ordered a few thousand freeze-dried toads so that they can study the use of their venom in traditional medicine. Perhaps there are cane toad pills in our future.

One final thought on the matter of cane toads: could the presence of a mind-altering substance in their skin have given rise to all those children's stories about the girl who kisses a toad and turns him into a prince? Perhaps. Who knows what you're going to see after you kiss one of those hallucinogenic warty-skinned amphibians.

Willow Power

It's decision time. I've been putting it off. But when you start receiving brochures addressed to you from the Golden Age Association — info on programs for the over-fifty set — you know the moment has come to get off the fence. While you're still nimble enough to do so, you've got to decide whether to start taking a small dose of aspirin on a regular basis in order to ward off some cardiovascular disaster.

Aspirin is a fascinating drug. It does so many things: it alleviates pain and reduces inflammation and fever; it may even prevent certain types of cancer; and, of course, it interferes with blood clotting. This latter effect was discovered in a serendipitous fashion by a California family physician way back in the 1940s. Dr. Lawrence Craven noticed that an unusual number of his patients developed bleeding complications after having their tonsils removed. And each of those patients had been chewing an aspirin-based gum to soothe their sore throats. Craven mulled this over. He knew that in many cases heart attacks were caused by the formation of blood clots that choked off the flow of blood in the coronary arteries, so he began to give aspirin to his patients who had suffered heart attacks. Within a few years Craven had proven that the incidence of a second heart attack was dramatically reduced by this method. But who would listen to a lowly general practitioner? Virtually no one. For about forty years his findings were essentially ignored.

Then, in the 1980s, several studies designed by high-powered researchers were released showing that aspirin could indeed lessen the chance of a second heart attack and even an initial episode. In one of these, The Physicians' Health Study, researchers gave either an aspirin tablet or a placebo to over twenty-two thousand male doctors. Among the aspirin takers they recorded such an astonishing reduction in heart attacks that they ended the study early to allow all study participants to start taking aspirin if they so desired. Other studies recorded equally dramatic reductions in strokes due to blood clots when subjects ingested aspirin on a daily basis. The results of these trials are so compelling that some researchers suggest as many as a couple of hundred thousand deaths from heart attack or stroke — in North America, among both men and women — could be prevented every year with the judicious use of aspirin.

That's quite a claim for the little pill that was concocted in 1897 by Felix Hoffmann, a chemist working for the Bayer company in Germany. While Hoffmann did synthesize the first commercial sample of acetylsalicylic acid, as aspirin is known generically, he wasn't the first to produce the substance in the laboratory. That honor goes to Karl Friedrich Gerhardt, who, in 1853 at Montpellier University in France, concocted an impure version with an eye towards improving on the effects of salicylic acid, a commonly used painkiller. At the time salicylic acid was extracted from the leaves of the meadowsweet plant and used for the treatment of fevers and pain, particularly of the arthritic variety. But it had to be taken in large amounts, it had a bitter taste, and it often caused stomach irritation. Gerhardt identified the molecular structure of salicylic acid and thought he could modify it to produce a better product. He abandoned the project when he found that he could not produce the acetylated version reliably.

How did we come to use salicylic acid as a pain reliever in the first place? It's a long and engaging tale that all starts with a recommendation inscribed in the famous 3,500-year-old Egyptian *Ebers Papyrus* to treat an inflamed wound with a concoction made from the leaves of the white willow tree. This makes sense in view of the fact that willow leaves and bark contain salicin, a substance that the body can convert to salicylic acid. Of course not all of the prescriptions for inflammation remedies in the *Ebers Papyrus* are as scientifically compelling — a poultice made from chopped bat and a potion of wasp's dung in fresh milk have not stood the test of time. But salicylate-containing plants have. Hippocrates championed the use of willow bark for easing the pain of childbirth, and the Roman physician Celsius described treating inflammation, characterized by redness, heat, pain, and swelling, with willow leaves. The ancient Chinese, as well as the North American Indians, knew about the special properties of plants like the meadowsweet.

The true scientific era of the salicylates began in England in 1763, when the Reverend Edward Stone presented a report to the Royal Society on the use of willow bark as a fever treatment. Stone was a believer in the rather curious Doctrine of Signatures, which maintained that one could find cures where the diseases themselves were spawned. Since fevers were often associated with swamps, probably because of mosquito-borne infectious agents, Stone searched swamplands for cures. He tasted a sprig of willow and was stunned by its bitterness. Aware that quinine, an equally bitter substance, was useful in the treatment of malarial fever, Stone decided to give willow bark a try. He dried the bark and powdered it and tried it on fifty patients with rheumatic symptoms. It worked.

The search was now on to discover the active ingredient. By 1828 scientists had isolated salicin, named after *Salix alba vulgaris*, the botanical name of the willow, and shown it to have a medicinal effect. Furthermore, they could convert it in the laboratory to salicylic acid, even more potent as a drug. At about this time Gerhardt became interested in solving the problems of bitterness and gastric complications, problems eventually solved by Felix Hoffmann some fifty years later. Hoffmann's father had long been taking salicylic acid for arthritis, but he could no longer consume it without vomiting. The chemist searched the literature for alternate forms of salicylates and came upon Gerhardt's work. By this time chemical techniques had been refined to the extent that Hoffmann was able to make acetylsalicylic acid in a pure form, thereby ushering in the aspirin era. The name was created by combining "spiric acid," as salicylic acid was originally known, with "a" for "acetyl."

While aspirin itself does not occur in nature, similar, less effective substances do. Willow extracts, sold in health food stores, cannot match aspirin's demonstrated effectiveness; in fact, aspirin came about as an improvement on the natural salicylates. Furthermore, we may attribute aspirin's anticoagulant effect to the acetyl part of the molecule responsible for deactivating an enzyme that leads to blood-clot formation. So there is really no point in chewing on willow bark to prevent a heart attack.

Recent evidence indicates that a dose of forty to eighty milligrams of aspirin a day is probably sufficient to bring about the cardioprotective effects. And at this dose the risk of gastric complications is very small. So is the risk of other complications, like the one experienced by a woman who took twelve tablets daily for her arthritis. She began to hear music, even when she sat in a soundproof room. When she reduced her dose to six tablets, the music stopped. Obviously, I'm not thinking about such huge doses — only a measly eighty-milligram tablet every day. I'm actually thinking hard. Especially in light of new studies that link aspirin with a reduced incidence of colon cancer, breast cancer, cataracts, and even Alzheimer's disease. Talk about music to one's ears.

Snake Oil for Aching Joints

According to the magazine article, Lucy was forty-eight when the pain of rheumatoid arthritis struck. She didn't take it lying down. The usual regimen of nonsteroidal anti-inflammatory drugs, cortisone, and methotrexate provided temporary relief, but the side effects were bothersome. So Lucy decided to go the alternative route. Whole hog. The first practitioner she saw told her to cut that out — the hog, that is. Not to mention beef, dairy products, potatoes, tomatoes, alcohol, sugar, wheat, citrus fruits, margarine, eggs, and all packaged and refined foods. Why? Because when Lucy had undergone applied kinesiology testing, she'd had a positive response to these foods. The therapist had measured the muscle strength in Lucy's arm after exposing her to the foods and determined that they were "too harsh for her and were sources of allergic reactions." If only allergy testing were truly that easy. But the elimination diet brought Lucy no relief, so she abandoned it and sought out other solutions.

At this point she began to explore an amazing array of treatments and dietary schemes. There were liver detoxifiers, intestinal detoxifiers, antiparasite formulas, live bacteria treatments, pancreas extracts, chelated minerals, and organic juice regimens. Then there were the herbal preparations, everything from devil's claw and celery extract to prickly ash and black currant seed oil. Lucy endured a battery of tests, including essential metabolics analysis, adrenal stress analysis, intestinal pH analysis, digestive stool analysis, and darkfield microscopy. She underwent acupuncture, tried homeopathy, fasted for days, ate clay, gulped algae, had her amalgam fillings removed, got vitamin B-12 shots, and took huge doses of folic acid and vitamin C. If there were a Guiness record for trying the most arthritis treatments, then Lucy would have set it.

While most of these treatments are on a very shaky scientific footing, desperate people still try them. History has shown that when scientific medicine leaves a vacuum, a host of alternative practitioners will rush in to fill it. These people claim to have the answers that have somehow eluded mainstream researchers. Such as the benefits of the Harmony Token, a colored disk you wear around your neck; its manufacturer claims that it "resupplies minerals, vitamins, and amino acids with the color that has been stripped away by exposure to electromagnetic radiation." Our bodies, apparently, do not recognize these colorless substances and, as a consequence, our immune systems are weakened. The Harmony Token utilizes 2,800 colors "to rebuild and repair the body at the cellular level and allows victims of rheumatoid arthritis to resume normal lives." Testimonials proclaim the disk's astonishing powers: it improves gas mileage and reduces car emissions; it makes racehorses run faster; it cures migraines. It also makes me wonder about people's sanity.

Lucy could have also tried the remedy described in an amazing book called *The Incredible Proven Natural Miracle Cure that Medical Science Has Never Revealed*. Let me reveal it to you: urine. Your own. All you have to do is take one to two drops daily, and feel your arthritis disappear. In extreme cases it has to be injected. If you're queasy about consuming urine, how about blood? Autohemotherapy involves the practitioner taking three-quarters of a cup of blood from a patient's vein, mixing it in a copper bowl with one-quarter of a cup of honey and one-quarter of a cup of lemon juice. Then the practitioner stirs and the patient drinks. There even seems to be some scientific backing here. *The Indian Journal of Orthopedics* reports that a majority of arthritis patients in one study experienced a reduction in pain and an increase in hand-grip strength. But maybe they were just afraid to admit that they hadn't improved lest they be subjected to more of the same treatment.

Then there is snake therapy, which is popular in some areas of China. The gall bladder of a living snake is removed and dropped into wine to make an arthritis cure. Most highly prized are the venomous snakes, like the king cobra. Apparently, these creatures sometimes escape from the cure-makers' shops, posing a threat to public safety. I guess seeing your mates being stretched out and cut open while they are still alive provides a certain motivation to get out of there.

This particular type of snake oil may be a hard sell in North America, but Jogging in a Jug has become a huge business. It was the brainchild of an Alabama farmer who was about to lose his farm and badly needed a money-making idea. It came in the form of his grandmother's arthritis remedy. Apple cider vinegar. He himself suffered from arthritis, and he'd found that his condition improved when he drank the stuff. When he began selling the concoction as an arthritis treatment, his financial situation improved, as well. At least until the U.S. government decided, based on the claims the farmer was making, that he was selling an unapproved new drug. The authorities destroyed thousands of bottles and required the farmer's company to send letters to its customers apprising them of the situation. Jogging in a Jug is not the only apple cider product for which manufacturers make miraculous claims. And those manufacturers all back up their claims with impressive-sounding testimonials. What they don't have, however, is scientific evidence. Such products cannot be taken seriously until they are studied seriously.

Researchers have attempted to do serious studies of some unusual arthritis treatments, such as the traditional raisins-in-gin remedy. The idea is to soak some raisins in gin for seven days and eat nine daily. A researcher at the University of North Texas has looked into this, and he maintains that it brings people with arthritis significant relief. He uses ninety-proof gin and has discovered that while soaking the raisins longer makes no difference, increasing the dose does. Maybe the remedy's effectiveness is due to anti-inflammatory compounds in the juniper berries from which the gin is made. Or maybe it's the alcohol. The researcher is now prescribing thirty-six gin-soaked raisins a day, and he insists that his arthritis sufferers feel much happier. I bet they do.

Some arthritis patients also say they feel better when they wear copper bracelets. Could there be something to this? An Australian study was done using anodized aluminum for placebo control. The researchers discovered that subjects wearing the copper bracelets did experience greater pain relief. By weighing the bracelets the researchers also learned that about thirteen milligrams of the copper dissolved per month; if the subjects' bodies absorbed all of this, then their copper levels would be higher than usual. Since many enzymes involved in tissue maintenance require copper to carry out their work, we can make a tenuous argument for copper bracelets. It is curious, though, that there is no evidence that oral copper supplements help arthritis sufferers.

We haven't yet reached the bottom of that huge barrel filled with alternative remedies for arthritis. We haven't broached aromatherapy, imagery ("Just picture the pain flowing out of your body and into the nearest creek . . ."), arnica poultices, Chinese Thunder God Vine (don't ask), or bee sting therapy (bee keepers who get stung an average of two thousand times a year are less prone to arthritis — or maybe they just don't notice it). Neither have we looked at borage seed oil, boron, ground ginger, boswelia, bromolain, reflexology, DMSO, or the Ayurvedic treatment that involves taking "yogaraj guggula" three times a day. We haven't looked at these because scientists haven't looked at them hard enough to come to any reasonable conclusions. And that, I'm afraid, is the case with so many of the arthritis treatments out there. They offer hope to people and often little else. Still, we keep looking.

Maybe the answer will come from gamma linolenic acid in evening primrose oil or from drinking orange juice laced with purified chicken cartilage. Maybe the secret lies in a blend of tea and cherry juice. Or maybe there is no secret, although Lucy would debate that. After three years of trying everything under the sun (including taking the sun each day), she says she's free of arthritis. Go figure.

Oil You Need to Know

Talk about milking a story. The newspapers got all excited when pop star Shania Twain revealed that the secret behind her soft, supple skin wasn't some complex, highly touted product filled with liposomes, collagen, or ceramides. The secret, she said, was udderly simple: Bag Balm delivered the goods. Just as it had done for cow udders since 1908, when a small Vermont company decided to take the bull by the horns and tackle the chronic problem of chapped cow udders.

Cow teats do not have an easy life. They are constantly being tugged, washed, disinfected, or wiped. This dries out the skin, which leads to cracking, which leads to infections, which leads to problems. Enter Bag Balm. Massaging the balm into the udder helps seal in moisture, prevents chapping, and reduces the occurrence of infection. The chemistry as it turns out, is pretty simple. All one needs to do is apply an antiseptic, provide a replacement for the natural skin oils that have been washed away, and coat the skin with a substance that prevents further moisture loss. One Bag Balm ingredient, 8-hydroxyquinoline sulfate, provides protection against bacteria and fungi, and another, lanolin, a fatty sheep skin secretion extracted from wool, reoils the udder. But the real key to Bag Balm's performance is the moisture barrier — good old petrolatum, perhaps more familiar as Vaseline.

When Bag Balm was first formulated, we already knew about the beneficial properties of Vaseline, thanks to the ingenuity of the American entrepreneur Robert Chesebrough. In the middle 1800s Chesebrough built a small company that distilled kerosene from coal tar. Then oil was discovered in Pennsylvania, and Chesebrough feared he'd be put out of business. Embracing the philosophy that if you can't beat 'em, join 'em, he traveled to the oil fields to investigate the possibility of producing his kerosene from this new source. One day as he walked through fields where pumps were busily bringing oil to the surface, Chesebrough's attention was attracted to a waxy material oozing out of the ground along with the petroleum. The oil workers hated the stuff because it gummed up their pumps. Nevertheless, they conceded that this "rod wax," as they called it, did have one asset: it helped heal cuts and burns!

Suddenly, the dollar signs began to flash in Chesebrough's head. After all, hadn't all kinds of people in the course of history anointed themselves with fats and oils for a variety of reasons? Hadn't his own grandmother smeared him with goose grease when he showed the first sign of coming down with a cold? So for the next ten years Robert Chesebrough devoted his energies to developing a pure, odor-free, effective form of rod wax — or petroleum jelly. He served as his own guinea pig, cutting his arms, burning his skin, pouring acids on his hands, and then treating his wounds with the jelly. The results were highly satisfactory. Chesebrough named his discovery Vaseline, supposedly because he used a vase of his wife's while conducting some early experiments. All that he had to do now was convince the public to buy it. So Chesebrough went out and became the pharmaceutical industry's first detail man. With a horse and buggy he cruised the streets of Brooklyn, New York, distributing free samples. The gimmick worked. Orders mushroomed and testimonials flooded in. Consumers discovered that Vaseline-coated metal didn't rust and that the stuff could protect long-distance swimmers in cold water. Vaseline could also soften leather, lubricate bedsprings, and, when smeared on the doorknob, keep children out of their parents' bedroom. But, most of all, people spoke about its near-miraculous effect on chapped skin.

When the producers of Bag Balm were searching for effective ingredients, petroleum jelly struck them as a natural. Cows were grateful. And, as it turned out, so were milkmaids. They noticed that their hands no longer chapped and that their skin felt silkier. The word was out. Bag Balm was in. Although the manufacturer has always maintained that the product is for veterinary use only, many people have purchased it over the years for their own purposes. While the business was not huge, it was vital enough to breed competitors with intriguing names like Udder Balm and Udderly Smooth. Although their manufacturers ostensibly aimed these products at the cow market, ads for them were replete with such terms as "greaseless feel" and "fresh fragrance." Hardly qualities of concern to cows, one would think.

Actually, chemists who develop cosmetics took up the challenge of cutting Vaseline's greasiness some time ago. Farmers weren't the only ones who recognized the moisture-retention properties of various petroleum derivatives; dermatologists have long known that nothing prevents skin dehydration better than Vaseline. But it is greasy, and how do you solve that problem? The obvious answer is to cut it with water. Yet, as we all know, oil and water do not mix. At least not without the help of emulsifiers. These molecules have a unique property: one end is attracted to water and the other to oil. When oil is shaken with an excess of water, it forms a suspension of tiny droplets that eventually coalesce, but when an emulsifier is present, it coats the droplets and prevents them from joining together.

Such an emulsion is an ideal moisturizing agent. It provides water for hydrating the skin and keeps the water from evaporating. Stearic acid, glyceryl stearate, and polyethylene glycol stearate are all commonly used emulsifiers. To improve the effectiveness of a moisturizing product manufacturers may mix in lanolin to replace the sebum lost from the skin through washing; they may also incorporate nongreasy synthetic oils like dimethicone. In other words, they can improve upon Vaseline. Indeed, Vaseline Intensive Care Dry Skin Formula, which makes use of this technology, always rates high in consumer-satisfaction surveys. Frankly, I find it more appealing than a product "intended primarily for the treatment of cow's udders" after "calving, high feeding, bruising, or chilling." Especially considering that makers of cow balm use lanolin, a substance that makers of hypoallergenic cosmetics have eliminated because of its irritant potential.

But I suppose the unusual has a certain appeal. So why not consider Crisco? Some plastic surgeons say it's the ideal material to aid the healing process after laser surgery for wrinkles.

It's a great moisture barrier. And if you don't like it, then you can always use the leftovers to bake a pie — if you're not worried about your intake of saturated fats, that is. Given the success of Bag Balm, perhaps I should also point out the existence of specialty products designed to eliminate the unsightly cracks that often sully the beauty of a pig's back. Pig Grooming Oil was developed by a large oil company to remedy this problem. I haven't yet heard any famous humans ascribing their lovely complexions to Pig Oil, but, hey — given the success of Bag Balm, there's probably a good business opportunity there somewhere.

Scents and Scents Ability

If you get a sniff of hexanal or (z)-4-heptenal, there's a good chance that someone's cooking an alligator nearby. These two unpleasant-smelling compounds form when alligator fat reacts with oxygen, particularly in the presence of heat. Alligator fat chemistry may not be your idea of a scintillating research topic, but the study of alligator-oxidation products is certainly important to alligator farmers in Florida and Louisiana. Alligator tail meat is low in fat and cholesterol and high in protein — well suited, therefore, for human consumption. It's too bad that it produces unappetizing off-odors when cooked; but researchers are becoming interested in how these compounds form and how they can be minimized. We now know that the smelly substances are oxidation products, so, to make alligator burgers a more attractive possibility, we may simply be able to add antioxidants to the meat.

Smell research extends way beyond malodorous alligators. Our lives are greatly influenced by fragrances, both in positive and negative ways. Just think about it. A sniff of a crayon can instantly transport us back to childhood. A whiff of pine may evoke a mountain vacation. The smell of cod liver oil still turns our stomach as it triggers memories of a parent forcing us to swallow the foul stuff. Indeed, our sense of smell is remarkable. We can detect thousands and thousands of different aromas and can even tell where they are coming from by the delayed arrival of the smell in the second nostril. This delay is only milliseconds long; nevertheless, it's discernible.

Besides inducing memories, smells can induce physical effects. The smell of an outhouse can make us gag, and bad breath can put an end to amorous feelings. The smell of a freshly baked pie can literally make us salivate. We find foods that smell good more appetizing. Can these observations be put to some use? Why not? We already use smells to affect other creatures — catnip to attract cats and naphthalene to repel moths — so why shouldn't we investigate the potential of fragrances to affect human life?

This is not a novel idea. We have long employed perfumes to add richness to our lives. As Kipling once astutely remarked, "smells are surer than sounds or sights to make the heart strings vibrate." We all accept that pleasant smells are attractive, but can we actually do more with smells? Like influence human health or behavior with specific fragrances? Believe it or not, researchers have done an impressive amount of work in this area, some of it whimsical, some deadly serious.

For example, a clever Japanese inventor has come up with a device that plugs into a person's nose where it releases smells like cocoa, bacon grease, or cheese when lettuce or tofu are eaten, making these bland foods more palatable. Scientists at the Smell and Taste Treatment and Research Foundation of Chicago have discovered that people are willing to pay more for Nike shoes in a store with a floral scent in the air. They also found that green-apple smell is effective against claustrophobia and that when a certain fragrance wafts through the air in a casino, slot-machine profits increase. Needless to say, the composition of this latter smell is a closely guarded secret, and nobody knows if any casino has actually tried to exploit it. But, as far as retailers are concerned, "fragrance planning" is already part of doing business. A children's clothing store in San Francisco has done well with cinnamon-and-hot-apple smell. It probably suggests to shoppers that the merchandise is as American as apple pie.

Supermarkets have played with the idea of boosting business in their bakery sections by spraying some of the compounds found in freshly baked bread into the air. Processed-food manufacturers have treated the packaging of their unpopped corn with popcorn smell to attract buyers. Banks have even experimented with coating their coupons advertising car loans with new-car aroma. A British company called Bodywise claims to have devised a way to make people pay their bills more promptly. They extract their product, Aeolus 7, from men's armpits and then apply it to outstanding bills. Bodywise conducted a trial in Australia and found that seventy percent more people paid their bills when the paper the bills were printed on was impregnated with the extract.

Another British company is bent on trying to make movies more exciting. It has patented a process that digitally encodes a film to trigger the release, at the appropriate moment, of smells stored in packets tucked into the theater's ventilation system. So far company officials haven't revealed what smells they're working on. Are we to be treated to the smell of horse dung in cowboy pictures, the smell of gasoline in car-chase scenes, or the fragrance of sweat in steamy love scenes?

French researchers have gone to the dogs and come up with a way to silence barking canines with smells. They developed a dog collar with a built-in microphone and a built-in supply of

citronella fragrance. When the dog barks, the collar sprays citronella under his nose. The animal then becomes preoccupied with figuring out where the smell is coming from and forgets to bark. Well, even if the collar doesn't silence your dog, mosquitoes won't bother him so much.

There is a more serious side to smell research. At the University of Vienna chemists tested the effect of the smell of lavender on mice. The creatures went to sleep and woke up when the smell was removed. This did not happen to human subjects, but they did become more relaxed. The Japanese found that this relaxation can increase the efficiency of computer operators, perhaps by preventing their minds from wandering. The workers they studied made twenty-one percent fewer errors when the workplace air was scented with lavender, thirty-three percent fewer with jasmine, and a whopping fifty-four percent fewer with lemon. Acting on such observations, one Japanese company designed a computerized system that sends fragrances through a building's ventilation ducts. Apparently, a lavender-chamomile mixture will reduce stress, while basil, peppermint, and clove fragrances will refresh workers. Cypress scent induces a positive mood — presumably, workers won't even think about going home.

In the United States, Yale University has taken out a patent on apple-spice fragrance because of its ability to stop panic attacks. In England, scientists at Warwick University have found that people's anxiety levels may drop by seventeen percent after they sniff a perfume that contains the essence of seaweed; it evokes the beach. But, according to researchers at the Monell Chemical Senses Center in Philadelphia, you don't even have to go to the beach to have your spirits lifted. You just have to smell an older lady's armpit. I kid you not. The Philadelphia research team asked university students to complete questionnaires about their mental state after being exposed to smells extracted from gauze pads worn by children, young adults, or senior citizens. These people had kept the pads in their armpits for ten hours. The smell of young men produced a depressive effect, while "old lady fragrance" had a decidedly uplifting one. Perhaps the students who participated in the study associated this smell with the comforting presence of a grandmother.

Obviously, smells can be pretty important. If you're trudging around the Everglades and become aware of a rancid, fishy fragrance, you'd better start looking out for alligators. The worse the smell, the larger your problem — big alligators produce a lot more (z)-4-heptenal than little ones. You may have to pull out that bottle of "l'eau de grandmere" and take a good whiff to calm yourself down.

The Man with the Silver Sock

Other patrons of the Gymtech Health Club, where I regularly pound the treadmill, may think I'm a little batty. That's because I've been working out while wearing a different sock on each foot. On the right I've got the standard-issue white cotton; on the left I'm going "x-static." That's the clever name given to a newfangled type of hose that claims to reduce foot odor in a completely safe fashion. And, to boot, it may even reduce the risk of athlete's foot. What's the secret? Silver.

The sock's manufacturers weave their product from nylon fiber coated with a thin layer of pure silver, a metal that has well-established antimicrobial properties. That's just what we need to control foot odor. The ghastly and pungent aroma arises when bacteria digest compounds found in sweat. One of these, butyric acid, can rival skunk fragrance in its ability to clear a room in a hurry. Undoubtedly, if it were not expressly forbidden by the Geneva Convention, someone would have already employed a teenager's ripened sneaker as a terrifying chemical weapon. While this, mercifully, has not yet happened, domestic squabbles have erupted at the mere threat of shoe removal. But now silver-impregnated fiber may restore domestic bliss.

We've long been aware that silver can effectively control the proliferation of bacteria on the foot — or elsewhere. Cyrus the Great made use of this fact way back in the fifth century B.C., when he ordered his servants to store his drinking water in silver vessels. Of course, Cyrus knew nothing about the chemistry of the situation, but the king had clearly paid heed to the stories passed down through generations about the health benefits of drinking water kept in silver containers. Legend also has it that Sir Lancelot won Guinevere's heart by a nose. His rival, King Arthur, favored golden armor, while Lancelot's armor was fashioned from silver. It was hot inside those suits, and the conditions were ripe for bacterial growth. Maybe the antimicrobial effect of silver was enough to make Lancelot more aromatically appealing.

Even our use of silverware may have an interesting health

connection. Throughout much of our history humankind hasn't paid a lot of attention to cleanliness. In earlier days improperly washed cutlery fostered the growth of disease-causing bacteria such as E. coli and salmonella. Silver spoons would have been less prone to this problem because of the metal's antibacterial effect. The rich could afford such cutlery and enjoyed the health benefits it bestowed; all of this contributed to the saying that the children of the wealthy were born with silver spoons in their mouths. Some Russian peasants still place silver coins in cisterns and jugs to keep their water fresh.

In North America we also make use of silver's power to disrupt microbes. Activated carbon water filters, for example, are sometimes impregnated with silver to discourage bacterial growth within the filter. And now we even know how this works. Metallic silver, due to contact with oxygen, is coated with a thin layer of silver oxide. Essentially, this means that silver atoms have donated an electron to oxygen and have become positively charged silver ions. It is these ions that bind to, and deactivate, various proteins that are essential to microbial life. Sometimes we use silver oxide directly. The makers of the silicone tubes that surgeons insert into the ears of children who are prone to ear infections impregnate their product with silver oxide. This reduces the risk of infection while the tubes are in place. Other silver compounds also have antimicrobial properties. Silver sulfadiazine is often used as a disinfectant for burns, and an eyedrop solution based on silver nitrate can ward off eye infections in newborns.

Obviously, silver has some useful medical properties. But could you be convinced that a colloidal solution of silver taken orally can combat chronic fatigue, meningitis, acne, arthritis, high blood pressure, syphilis, shingles, pneumonia, AIDS, and cancer? I hope you couldn't. But that is exactly what some purveyors of "colloidal silver" products (a very fine suspension of silver particles stabilized with gelatin) are trying to do. One advertiser unabashedly proclaims, "You can live disease-free, add countless years to your life. Colloidal silver is the greatest disease fighter known to man. It will kill by suffocating over 650 bacteria and viruses." Where do they get this stuff? It is true that silver can kill a large variety of microbes in a test tube, but this in no way proves that the metal prevents disease in humans who ingest it. Those hucksters are clever, though. They refer to the scientific literature selectively and use it to guide the gullible to the wrong conclusions.

They badger us with tales of how doctors used silver nitrate to treat infections before the large pharmaceutical companies got into the game with their patented antibiotics, which generated greater profit. Now, they claim, we are paying the price: antibiotic-resistant bacteria. We should have stuck to "natural" silver. Furthermore, our crops are now grown in silver-depleted soil and all kinds of diseases are on the rise due to a silver deficiency in our diets. What balderdash! Yes, we did once use silver nitrate as a disinfectant. Babe Ruth was even treated with it for his throat problem; it only made it worse. What the colloidal-silver-tongued orators fail to mention is that silver is not an essential nutrient and that silver nitrate fell out of favor with physicians because it didn't work very well in overcoming infections. That's not the only reason doctors gave it up, though. They also didn't like the fact that the stuff sometimes colored their patients' skin gray — permanently. Argyria, as the condition is known, is irreversible. Some victims of the inappropriate use of silver compounds during the forties and fifties are still around to serve as living testimonials to this particular medical folly.

The U.S. Food and Drug Administration has clamped down on colloidal-silver preparations, banning the sale of any such product that claims to deliver health benefits. Unfortunately, colloidal silver can still be sold as a dietary supplement, as long as manufacturers make no health promises to the consumer. In the legislative context dietary supplements constitute a gray area — in the case of colloidal silver, perhaps quite literally. Leave it to those ingenious hucksters, though, to find ways to compensate for any loss of sales. Now they've taken to selling an apparatus you can use to make colloidal silver at home. All you have to do is pass an electric current through a solution equipped with a silver electrode. Actually, the only thing this device will accomplish is to line the promotors' pockets with silver. The whole thing stinks. But at least my left foot doesn't.

That's right. The sock really did the job. How do I know? Because to analyze my research I used the most sophisticated smell-detecting instrument available to humankind: my wife's nose. Believe me, nothing gets by it. The verdict was clear. While she was not exactly thrilled about performing the task, she did determine that the X-static sock was up to snuff. That was enough to start me thinking about other possible applications. Perhaps silver-coated fiber is the right stuff to control the classic underarm bouquet. But this will need testing, so don't be surprised if you see me running on the treadmill with an X-static sock tucked in my armpit.

Mercury: Mysterious and Deadly

Not long ago I tracked down some research papers authored by Karen Wetterhahn, and I was amazed by their diversity. Wetterhahn, the first female professor of chemistry ever hired at Dartmouth, was interested in fields ranging from biochemistry to toxicology. She was an expert on the cancer-causing potential of chromium, and, like numerous chemists before her, she was fascinated with the chemistry of mercury and its compounds.

The Ancient Greeks were well aware of the properties of this liquid metal. They could make a glob of it scatter into tiny droplets so quickly (hence the term "quicksilver") that they named it after Mercury, the fleet messenger of the gods. The Chinese were also intrigued by the appearance of mercury and thought that it held the key to longevity and perhaps even immortality. They were wrong.

In fact, the Romans had already noted that the men who mined cinnabar (mercuric sulfide), the naturally occurring ore from which mercury is extracted, had a short life expectancy. Accordingly, they sent only the worst criminals off to toil in the quicksilver mines of Spain. Some of these men would suffer from tremors, excessive salivation, fits of hostility, and memory loss. Indeed, the outward beauty of mercury belies its extreme toxicity.

The alchemists of the Middle Ages learned about this the hard way. They believed that all metals were mixtures of mercury and other substances, and they struggled to combine mercury with sulfur to make gold. Instead of wealth, they acquired mercury poisoning. The great Isaac Newton may well have been among them. Most people connect Newton with falling apples, but the man also dabbled in alchemy. In his notebooks he recorded numerous mercury experiments and described his concoctions as being "sweetish," "saltish," and "vitriolic." He obviously tasted his products, so we should not be too surprised that in 1692 he started to betray signs of mental illness. Newton suffered delusions, periods of melancholy, sleeplessness, and loss of appetite — all typical symptoms of mercury poisoning. When Newton left his laboratory to become Master of the Mint in London, the symptoms disappeared and his brilliance returned.

But not everyone exposed to mercury was so lucky. Some were driven mad by the lustrous metal. Hatters, for example.

Lewis Carroll's fictional Mad Hatter could well have been based upon some real people. In those days the felt used for hatmaking was made from rabbit, hare, or beaver fur, which the hatter would mat by treating it with hot mercuric nitrate. Erratic behavior, along with swollen gums and tremors, became the recognized consequences of hatting.

Today we no longer tinker with alchemy, and mercury has long been abandoned by hat manufacturers. But this doesn't mean that we don't still have to contend with mercury exposure. The chemical industry employs the metal as a catalyst, and mercury is an integral part of the process used to produce chlorine and lye (sodium hydroxide) from salt. It is also found in electrical switches, dental amalgams, batteries, detonators, and, of course, thermometers.

One would not think that a shiny little sliver of mercury sealed in a thermometer could cause problems, but broken thermometers have resulted in a number of hospitalizations. Children are especially susceptible to the effects of mercury vapor, because of their small size. The symptoms they experience may include loss of appetite, drooling, excessive perspiration, listlessness, itchiness, and swollen, reddish, cold extremities. Rapid heart beat and elevated blood pressure may also occur.

Unfortunately, mercury poisoning is not always the first thing that comes to a physician's mind when he or she is confronted with such symptoms. A proper diagnosis may take some time. Clearly, prevention is the best protection. We should clean up a mercury spill immediately. Vacuuming is generally not the best cleaning method, because a vacuum cleaner will separate the mercury into tiny droplets and spread them throughout the air via the exhaust; but if the spill is on the carpet, there is really no alternative. Remove the used bag from the vacuum cleaner right away, place it in a plastic bag, and discard it as toxic waste. Most municipalities have special arrangements for disposing of such waste.

As dangerous as mercury vapor is, it is by no means the most toxic form of the metal. This distinction belongs to a compound of mercury, dimethyl mercury. The world found out about the deadly nature of this substance in the 1950s, when dozens of people died and thousands experienced symptoms of mercury toxicity in the Japanese fishing village of Minamata. A nearby chemical company had been using mercury in the production of acetylene, and it had routinely discharged the used mercury into the ocean. Mercury isn't soluble in water, and it should have simply accumulated harmlessly at the bottom of the sea. But it didn't.

Microorganisms in the water converted the insoluble mercury to soluble dimethyl mercury, which then contaminated the fish. The first sign of trouble was an outbreak of "cat-dancing disease." Many of Minamata's felines staggered, salivated, convulsed, and collapsed after eating a meal of fish. Soon humans began to follow suit. The final chapter in this tragedy remains unwritten, since studies are now showing that children born to women who were exposed to the dimethyl mercury are more prone to neurological diseases such as cerebral palsy.

There are various theories about how mercury causes its terrible effects. In all likelihood it binds to, and deactivates, critical enzymes. In order to do this, the toxin must enter the bloodstream. As a vapor, mercury can accomplish this easily; it can also do so as a soluble compound like dimethyl mercury. But swallowed mercury metal, because it lacks solubility, is far less dangerous. So is the mercury in dental amalgams, which is so tightly bound that only extremely tiny amounts are released into the bodies of those who have them. There are people who argue that even these trace levels can cause various toxic symptoms, but in the scientific community an overwhelming majority concurs that amalgam fillings are safe. What remains a mystery, however, is exactly how mercury interferes with the transmission of genetic information.

Professor Wetterhahn addressed this very question in 1997 by conducting an experiment that involved adding a tiny amount of dimethyl mercury to a sample. Working in a fume hood, wearing latex gloves, Wetterhahn carefully added the liquid. Then Murphy's Law kicked in, and she accidentally spilled a couple of drops on her gloved hand. She performed a quick cleanup and went back to what she was doing.

Wetterhahn's work was incredibly varied. Besides teaching and doing research she served as dean of the Faculty of Sciences and was cofounder of Dartmouth's Women in Science Project. When she took time off, she would golf, garden, or listen to Bruce Springsteen albums. Three months after the dimethyl mercury spill Wetterhahn began to experience episodes of nausea and vomiting. Five months after the accident her speech became slurred, her vision and hearing began to fail, and she could no longer walk straight. A short time later she lapsed into a coma and died. She was only forty-eight years old. A couple of drops of dimethyl mercury had taken a mother from her children, a wife from her husband, a friend from her community, a bright light from science.

I never knew Professor Wetterhahn, but her death moved me deeply. Primarily, of course, what happened to her was a personal disaster for her and her family. But it was also a demonstration of the power of the scientific spirit — that willingness to take calculated risks for the benefit of others. In a sense Karen Wetterhahn died in the line of duty, in the pursuit of scientific knowledge. She will be missed and remembered.

FOOD MATTERS

You Are What You Eat

The Imperial Herbal Restaurant in Singapore surely is a fascinating place. A meal can cost you over five hundred dollars, but for that sum you not only get a dining experience, but you also get a dose of health. Or so the faithful claim. The restaurant serves a Lingzhi herbal soup, which prevents cancer, heart disease, insomnia, and asthma. You may follow this with an appetizer of drunken scorpions with asparagus, said to be very soothing to the nerves. Not a bad idea, since some of the items on the menu can be quite unnerving. Like a general pick-me-up consisting of snow-frog glands topped with stewed fruit bat accompanied by a potato sprinkled with fried black ants. And if you have a more specific complaint, a Chinese apothecary is available for consultation. Depending on your problem, he may advise you to order up some dried locusts, a few gecko lizards, or a soup made from the desiccated reproductive organ of the male deer. The latter is somewhat pricey, at $425 a bowl, but its promise is greater than that of Viagra. Customers on a budget may order a soup made from the corresponding organ of a bull, instead.

We may snicker at these delicacies and the health claims that go with them, but the fact is that food was probably humankind's first medicine. After all, when our primitive ancestors felt sick, what else could they do but manipulate their diets? In some cases they hit upon remedies that worked. The Egyptians prescribed extract of pomegranate for intestinal worms and roasted liver of ox for night blindness. Pomegranates actually contain a vermifuge, and the high vitamin A content of liver could have countered visual problems caused by a deficiency of the vitamin.

Given such empirical evidence, it isn't hard for us to see why even ancient civilizations held to the you-are-what-you-eat principle. However, they did sometimes push it to ridiculous extents and would consume, for example, lion's heart for courage or tiger's bone for strength. Today the idea that food governs health remains prevalent, and much sophisticated science is founded on it, the Imperial Herbal Restaurant notwithstanding. We seek out broccoli for its anticarcinogenic effect, tomatoes for their antioxidants, soybeans for their natural estrogens, and whole grains for their vitamins and fiber.

Although treating illness with food is an ancient notion, the idea that diet can prevent disease only dates back to the last century. The first health reformer to make extensive use of diet was the Presbyterian minister Sylvester Graham. His is an amazing story. Virtually everyone has heard of the riot sparked by the American colonists, dressed as Indians, who threw tea into the Boston harbor, but not many people know that in 1837 Boston experienced the notorious Graham riot. This time colonists dumping tea didn't cause the uproar; it was bakers dumping whole-grain flour over people waiting to hear a lecture by Sylvester Graham.

During the 1800s traveling lecturers were a major source of diversion for an entertainment-starved public. These speakers, quasi-entertainers, pontificated on matters ranging from religion to nutrition, often with a fair degree of overlap. Sylvester Graham was the most famous of these itinerant lecturers. The seventeenth child of a Presbyterian minister, young Sylvester also took on the challenge of becoming a man of the cloth. His sermons on temperance were so popular that he was in constant demand. Convinced that proper eating was essential to good health, he added food to his list of lecture topics. "Every affection and every disturbance of the stomach influences, in a greater or lesser degree, every organ and every function of the body," he claimed. Pretty advanced thinking for the 1800s, an era in which many guzzled whiskey for breakfast and slathered lard on bread for a snack. Graham frequented medical libraries and often tried to buttress his arguments with his interpretation of the tenets of physiology. He unabashedly called his lecture series "The Science of Human Life."

Graham's lectures were undoubtedly interesting and entertaining, but they weren't altogether scientific. Much like the diatribes of today's health gurus, they were laced with personal anecdotes and testimonials. Graham's pat remedy for physical and moral decay was to increase consumption of fruits and vegetables and avoid meat. He linked the eating of flesh to what he called the "great evil" — namely, masturbation. It was one of the roots of all disease, he insisted. Even orthodox sexual activity earned Graham's scorn. Too much sex, or even thinking about sex, predisposed an individual to disease, but a person could curtail unruly passions by eating fruits, vegetables, and whole-grain bread. Graham prohibited salt and was particularly venomous in his attacks on chicken pie, which he said caused cholera.

Grahamite rooming houses sprang up, where boarders could thrive on meatless meals consisting of coarse whole-wheat bread, uncooked vegetables, oatmeal, and barley — not unlike the meals that today's nutritionists are encouraging us to consume. Yet Graham hadn't a clue about vitamins, antioxidants, or fiber. He was just trying to devise a diet to curb the sexual appetite.

Wailing relentlessly about the horrors of white bread, Graham angered the bakers of the period. Bread made from refined white flour was their staple, the product upon which the industry was based. They picketed Graham's lectures, and their opposition to his ideas culminated in the Graham riot of 1837. The bakers achieved their purpose by preventing the preacher from delivering a widely advertised lecture on the benefits of fresh air, exercise, bran consumption, and meatless and alcohol-free meals. Too bad, because if more people had listened to Sylvester Graham back then we may never have reached the frightening levels of heart disease we saw during the first half of this century.

Sylvester Graham is all but forgotten these days, but his legacy does remain in the form of Graham flour, which is a coarsely ground whole-grain product. And, of course, we have the Graham cracker, but Graham did not create it. The original Graham cracker was a chewy, coarse, whole-wheat bread first sold by James Caleb Jackson in the 1860s. Jackson wanted to create a healthy cracker that would keep indefinitely to sell to his patients at his water-cure establishment, Our Home on the Hillside, near Dansville, New York. Jackson mixed Graham flour with water, baked it in sheets, broke it up, and baked it again. The crackers became so popular that Nabisco began to market them commercially in 1898. There was nary a mention of Graham's claim that whole-grain products would "reduce the passions."

FEELING HOT, HOT, HOT

Let me make a case against Capital Punishment. I tried it and I didn't like it. I would venture to say that few among you would. It was rated at one hundred thousand Scoville units, which, you should know, places this hot-pepper sauce roughly in the same heat category as glowing charcoal. Chomping on glowing charcoal may actually have been more pleasant.

Way back in 1912 pharmacist Wilbur Scoville figured that the world needed a way to measure the spiciness of peppers, so he devised the Scoville Organoleptic Scale. To do this he had to assemble a panel of brave judges, whom Scoville instructed to sample increasingly diluted extracts of hot peppers until the flavor became undetectable. The hottest pepper they tested, the habanero, came in between two hundred thousand and three hundred thousand Scovilles, cayenne at thirty-five thousand, and the popular jalapeno at about four thousand. The compounds responsible for the heat of peppers are the capsaicinoids, particularly one called capsaicin. Pure capsaicin rates a mind- and mouth-numbing sixteen million Scoville units. The judges who determined this rating undoubtedly had a memorable experience.

The chemistry of such an experience is now quite well understood. Capsaicin interacts with a specific protein, a "receptor," on the surface of nerve cells and triggers an influx of calcium into the cell. This then liberates a string of amino acids, known as "substance P," from nerve endings. These send the pain message to the brain. And that pain can be intense.

Pity the young man who burst into a Chicago clinic, waving his hands and moaning in pain. With some difficulty he described to the attending physician how he had been in the midst of preparing hot peppers for a Hunan Chinese lunch when he began to experience a terrible burning sensation in his fingertips. The pain radiated up his arm, his face turned red, he perspired profusely, and he started to feel faint. Then the doctor learned that just prior to his medical emergency the man had been sanding furniture. Now the doctor saw the sequence of events clearly. The unfortunate gentleman's fingertips were abraded by the sanding, allowing the capsaicin from the peppers to be absorbed directly into his bloodstream. Treatment with a local anesthetic cream brought him relief from Hunan hand.

If you think Hunan hand sounds painful, imagine the torture of jalapeno eye. This time the victim was a backyard gardener who had just harvested her crop of jalapeno peppers. She washed her hands and then proceeded to put in her contacts. She immediately felt a piercing, burning sensation in her eyes. This lady can now identify with the victims of pepper spray, the special capsaicin preparation used by police to subdue rioters and violent criminals. It works. It's tough to continue rioting with cayenne pepper extract on your face. But some worry that in rare cases the effect may be stronger than desired. Like death. When experts examined cases of fatal capsaicin overdose, they concluded that some people, particularly if they suffer from mental illness or are under the influence of drugs, may not be stopped by the usual dose of pepper spray. Police officers may then resort to excessive use.

You wouldn't worry about excessive use, though, if you found yourself in the Alaskan wilderness gazing into the eyes of a charging bear. You would quickly reach for your Bear Guard pepper spray, guaranteed to stop that bear in its tracks. Would it actually work? Nobody really knows. The American Environmental Protection Agency discourages people from using such products because the manufacturers have been unable to submit data proving that their products are effective. No wonder, considering what they'd have to do to obtain that data.

If we can't repel bears with capsaicin, then we can use it to send squirrels scurrying. They hate hot peppers, but, to the great annoyance of bird fanciers, they love bird seed. Birds have no receptors for capsaicin and are therefore immune to its effects, so why not treat bird seed with capsaicin to keep the rodents away? Such products are actually being developed under the Squirrel Free brand, and they are pretty hot. You need a hit of about twenty thousand Scoville units to get squirrels to hightail it, and this requires about thirty dried habaneros per pound of bird seed.

Habaneros can also spice up paint. This was the brainstorm of an American sailor named Ken Fischer. One day, while eating a deviled egg topped with tabasco sauce, Fischer had a devil of an idea. Boaters are constantly plagued by barnacles, tubeworms, and zebra mussels, which adhere to the hulls of their vessels and increase drag. The fire in Fischer's mouth made him wonder if barnacles and their ilk could be repelled by the same sensation. With the help of the McCormick spice company, Fischer came up with Barnacle Ban, a substance that is more environmentally friendly than the toxic, metal-based antifouling paints boaters generally use. Barnacle Ban doesn't kill the barnacles, it just encourages them to move on to a "cooler" climate.

If we can discourage sea creatures with capsaicin, why can't we use it to keep insects at bay? Fruits and vegetables are extremely prone to insect infestation, so pesticides that are nontoxic to humans would be most welcome. Enter Hot Pepper Wax. If we spray infested crops with this wax, then seventy percent of the insects on them die due to overstimulation of the nervous system. The capsaicin causes them to defecate endlessly until they die. Lovely. In humans such overstimulation can have a very different effect: it can actually kill pain, such as the excruciating variety associated with shingles.

The virus that is responsible for chicken pox also causes shingles. After the symptoms of chicken pox disappear, the varicella-zoster virus can take up residence in the nervous system, where it lies dormant. Many years later, usually because of a weakening of the immune system due to age or disease, the virus may become active. The result now is not chicken pox, but shingles. The name derives from the Latin *cingulum*, meaning "belt," because the blisters and rash that are the hallmarks of the disease usually appear around the waist. Other areas of the body, such as the face, can also be involved. The symptoms usually disappear after a few weeks, but in rare cases the pain can persist for months, or even years. Postherpetic neuralgia can make the skin so sensitive that the person suffering from the condition can't even tolerate clothing. Now there is some hope for the management of this horrendous pain: a cream formulated with capsaicin. After triggering an initial release of substance P, the capsaicin in the cream prevents nerve cells from reaccumulating it. Still, the medication is not problem-free. As one would expect from a product formulated with hot peppers, the cream can produce a burning sensation on the skin, although this usually subsides with prolonged application.

Researchers have also issued encouraging reports about capsaicin-based creams as a treatment for arthritis and the neurological pain associated with diabetes. Now we can even buy capsaicin poultices for lower back pain. The compound appears to be a very safe medication; in fact, when researchers used a gastroscope to examine the stomachs of subjects who had ingested the spicy substance, they saw no signs of irritation. And wait, there's more. Capsaicin has anticoagulant properties and may reduce the risk of stroke and heart disease. It even appears to prevent carcinogens from binding to DNA, and we may therefore refer to it as an anticarcinogen. Furthermore, an Italian study has shown that cluster headaches may be alleviated by spraying capsaicin up the nostril on the side of the face where the headaches occur. Based on my experience with Capital Punishment, I don't know about that one. And what can you do if you have been Capitally Punished? Water is useless, since capsaicin doesn't dissolve in it. But a liquid containing some fat, like whole milk, will do. The best solvent for capsaicin, though, is alcohol. So your final request should be a margarita.

The Evolution of Margarine

Who would have ever thought that we would one day drug ourselves with margarine? Certainly not Hippolyte Mege-Mouries, the French chemist who first developed the spread in the 1860s. But, then again, his margarine was nothing like the "nutraceutical" blends that are cropping up on grocery store shelves around the world. These products have special ingredients that will do more than merely provide basic nutrition.

They will, for example, lower our blood-cholesterol levels. Believe it or not, margarine manufacturers have the scientific studies to back such claims.

First things first. What was Mege-Mouries thinking of when he first concocted margarine? In a word, money. Napoleon III had offered a prize to anyone who could create a cheap butter substitute, and Mege-Mouries felt inspired. He knew that butter was essentially milk fat, but he began to wonder where the fat came from. Since milk contained fat even when cows were undernourished and were losing weight, he concluded that milk fat came from the cow's body fat. So the inventive chemist chopped up some suet, poured in some milk, added a little minced sheep stomach for texture, and cooked the mixture in slightly alkaline water to get "butter." The concoction looked like butter, but it didn't taste very good. It didn't have enough "cow" flavor. Mege-Mouries's solution to this was to add some chopped cow udder. That apparently did the trick, because in 1870 Napoleon III awarded him the prize and gave him a factory in which he could mass produce the new product. All that Mege-Mouries needed to do now was find a name for the spread.

Earlier, in 1813, another French chemist, Michel Chevreul, had isolated an acidic substance from animal fat that formed intriguing pearly drops. He named it "margaric acid," from the Greek *margaron*, or "pearl." Margaric acid later turned out to be one of the fatty acids commonly found in animal fat. Fats, also referred to as triglycerides, have a molecular structure that resembles a comb with three teeth. The teeth are the fatty acids, and the spine to which they are attached is a three-carbon molecule called glycerol. These fatty acids are composed of chains of carbon atoms of varying lengths, so numerous triglycerides are possible. By the time Mege-Mouries made his discovery, chemists already knew that margaric acid had seventeen carbon atoms in its chain. We know this from a most unusual source: Victor Hugo. His epic work, *Les Misérables*, written in 1862, contains the arresting passage, "Comrades, we will overthrow the government, as sure as there are fifteen acids intermediate between margaric acid and formic acid." Indeed, that's exactly how many there are. Formic acid, with its one carbon atom, was first extracted from ants. Not surprising — ants are a lot smaller than cows.

In 1873 Mege-Mouries secured an American patent for his invention, but he never enjoyed the fruits of his labor. Margarine was not an instant success, and its inventor died in obscurity. But leave it to American ingenuity: by 1881 improvements to the original formulation resulted in a more appealing product. Manufacturers churned melted fat with milk and salt, chilled the mixture, and then kneaded it to a plastic consistency. This "new and improved" margarine sold well enough to provoke the ire of the dairy industry, which regarded this butter imposter as a potential competitor. The dairy lobby swung into action and buttered up enough congressmen to get the Margarine Act of 1886 passed. The government imposed a tax of two cents per pound on margarine; it also forced manufacturers, wholesalers, and retailers to pay licensing fees. In spite of this, sales increased, especially after 1905, when the process of hydrogenation was introduced. This process hardened liquid vegetable oils to the consistency of animal fat by reacting them with hydrogen gas. Margarine makers could now replace animal products with cheap vegetable oils.

The more margarine people ate, the more worried the dairy industry became. Once again it exercised its clout, spurring legislation to equalize butter and margarine prices and to prohibit the sale of margarine that was colored to make it look like butter. But the government began to ease restriction in the 1930s, after margarine manufacturers learned to use domestically produced cottonseed and soybean oil instead of expensive imported oils. American farmers, happy to have a vast new market for their crops, backed the initiative. Still, the butter-margarine skirmishes continued. In fact, in the 1940s they took on a new dimension. When studies investigating the detrimental effects of animal fats and cholesterol on cardiovascular health began to accumulate, margarine found a new calling. It had lucked into a new career as a "health food." Since it contained no animal products, it had no cholesterol, that cardiovascular demon. People switched from butter to margarine, hoping to protect their arteries from the horrors of cholesterol buildup.

These days switching is no longer the issue. People are concerned with choosing the right type of margarine, the type made with either sitostanol (Benecol) or sitosterol (Take Control). These chemicals may not sound particularly appetizing, but I suspect that a lot of people will eventually eat them. Just like the Finns, who have been stuffing themselves with Benecol margarine since 1995. Why? Because Finland has one of the highest rates of heart disease in the world, a rate that sitostanol ingestion promises to reduce by lowering blood-cholesterol levels. Here's the lowdown.

Finland has a huge forestry industry, and it sponsors a great deal of scientific research involving its own by-products. One of these is sitostanol, derived from pine wood pulp. This compound initially aroused interest because its molecular structure is similar to that of cholesterol. In the course of experiments designed to explore the potential use of pine-tree products in animal feed, researchers noted that the blood-cholesterol levels of animals that consumed sitostanol decreased. Human studies conducted shortly afterwards showed that about three grams of sitostanol per day reduced blood cholesterol by ten to fifteen percent, a result comparable to that for cholesterol-lowering medications. Excited by these results, researchers began to explore ways of incorporating sitostanol into the diet. Why not margarine? they wondered — people already eat the stuff in an attempt to reduce cholesterol. But, unfortunately, sitostanol was not soluble enough in fat. Chemists quickly solved this dilemma by converting sitostanol into sitostanol ester, which could easily be mixed into margarine.

Scientists theorize that, due to its chemical similarity, sitostanol competes with cholesterol to be absorbed into the bloodstream from the intestine. This not only interferes with the uptake of cholesterol from the diet, but it also, more importantly, lowers the amount of cholesterol that wends its way into the blood from cholesterol synthesis in the liver. Cholesterol is an essential biochemical, which the liver can supply, but much of it accesses the bloodstream in an indirect fashion. First it's secreted through bile into the intestine, where it plays a role in fat absorption, and then it is absorbed into the blood. Sitostanol blocks this absorption.

Both Benecol and Take Control (which contains a similar compound isolated from soybeans) have been approved for sale in the United States. We require two tablespoons per day to obtain the cholesterol-lowering effect. For how long? For as long as you want to experience the effect. Stop the margarine and your cholesterol will rise. This makes Benecol and Take Control manufacturers happy, especially since these margarines cost about five times as much as regular margarines. We don't yet know what all the long-term effects of eating these products are, but the cholesterol-lowering effect is quite real. Finally, how do these newfangled margarines taste? Not bad. If only they could find a way to make them taste more like butter. Someone should offer a prize.

A Different Twist on Licorice

Many consider *The Gold Rush* to be Charlie Chaplin's greatest movie, probably because of the hilarious shoe-eating scene. In this 1925 epic the Little Tramp travels to the Klondike in search of gold and finds a heap of trouble instead. Trapped in a cabin by a blizzard, he soon runs out of food. What can he do but boil his shoe? We see him dining on the sole, looking as satisfied as he would if he were eating a piece of steak. And I really mean dining, not just pretending. How did Chaplin do this? Simple. The special prop was made by the American Licorice Company, and it was undoubtedly far tastier than shoe leather. But I'm sure that Chaplin never considered the possible health consequences of his meal.

Now that I've raised the subject of licorice, I'm sure many people's thoughts will turn to the movies. It won't be Chaplin's culinary escapades that they think of, though — it will be the red or black stringy stuff that competes with popcorn at virtually every theater lobby snack counter. Actually, the red version has nothing to do with real licorice, and the black has only a very tenuous connection. The licorice plant has a long and fascinating history. Four-thousand-year-old Egyptian hieroglyphics describe how people used it as a medicine, and the ancient Greeks employed the plant's roots as a sweetening agent. Indeed, our word "licorice" evolved from the Greek word *glykyrrhiza*, meaning "sweet root." Nobody knows who first had the bright idea of boiling the root in water and then heating the resulting solution until the water evaporated to produce the familiar black mass we call licorice, but it was probably an attempt to produce a medicinal substance. Botanical history is filled with tales of our attempts to capitalize on the curative properties of natural substances. And the properties of licorice root would certainly have encouraged experimentation.

Various ancient manuscripts contain instructions for treating coughs and digestive problems with the root.

Licorice is one of the most widely investigated plant products. Modern chemistry has allowed for the isolation, separation, and characterization of dozens of different compounds found in the root extract. No single component accounts for licorice's distinctive flavor, but anethole comes close. This compound occurs in the anise plant, from which it can be extracted, or it can be synthesized in the laboratory. Candy makers commonly use anethole to impart a licorice flavor to their products, like that black stringy confection. This is why naming the treat "licorice" is misleading. Even when it is made with real licorice extract, the concentration of licorice compounds is very low. We should be aware of this, because it means that neither the problems nor the potential therapeutic effects that we attribute to real licorice apply to the twists we chew on in movie theaters. Real licorice twists do exist, though — mostly in Europe.

What problems and what therapeutic effects, you ask? Let's start with the problems, like the calamity that befell a twenty-year-old woman who had to be admitted to hospital because she had lost all strength in the lower half of her body. Blood tests revealed that her potassium level was extremely low. An astute physician asked her about her dietary habits, only to discover that the young lady was virtually addicted to licorice candies. She had been eating up to half a pound a day. Problem solved. The most prevalent compound in licorice, and the most studied, is glycyrrhizin, also known as glycyrrhizic acid. This substance has hormonal effects resembling those of aldosterone, an adrenal-gland hormone responsible for maintaining mineral balance in the blood. It helps the body retain sodium and excrete potassium. An excess of aldosterone, or compounds that behave like it, will cause excessive sodium retention, which, in turn, causes excessive water retention, which then causes high blood pressure. Loss of potassium can affect nerve and muscle function. The licorice-munching patient's doctor prescribed a potassium supplement, and it rapidly set things right.

This case is not unique. A man who switched to a licorice-flavored beverage when his doctor ordered him to give up alcohol also ended up in hospital suffering from weakness, high blood pressure, and low potassium. So did a man who had managed to give up smoking by switching to chewing gum flavored with real licorice. His severe abdominal pains turned out to be attributable to potassium loss. A man who chewed about ten three-ounce bags of tobacco a day became so weak he couldn't raise his arms; licorice is often used to flavor tobacco. In fact, we use about ninety percent of the licorice imported into North America for this purpose. When volunteers were asked to eat one to two hundred grams of licorice candies daily, they showed serious symptoms within a few weeks. The lesson here is obvious: do not consume unusual amounts of authentic licorice, especially if you have a history of high blood pressure or other medical conditions, like diabetes, heart disease, or glaucoma.

Why should anyone consider eating exceptional amounts of licorice, anyway? Consulting the herbal or alternative literature, one finds plenty of reasons. Alternative publications often tout licorice as a "natural" treatment for conditions ranging from colds and prostate problems to indigestion and cancer. Also, they frequently blow the results of animal experiments out of proportion and exaggerate human experiences. For example, just because mice exposed to a carcinogen developed fewer tumors when they drank water laced with glycyrrhizin does not mean that licorice is an effective treatment for human cancer. Just because there is some evidence that licorice can help to heal peptic ulcers in humans does not mean that it is the best therapy. It may simply have been an appropriate treatment until newer, far more effective prescription drugs came on the scene.

One area of investigation, however, does merit a closer look. And that, believe it or not, is the use of licorice in the treatment of chronic fatigue syndrome (CFS). Sufferers of this baffling and potentially devastating condition experience headaches, muscle pains, "mental fog," and depression. In 1995 Dr. Riccardo Baschetti, a retired Italian physician, submitted a letter to *The New Zealand Journal of Medicine* describing how he had cured his own CFS with licorice root. Because he felt better after eating salty foods, Baschetti wondered whether licorice, which he knew caused sodium retention, might be an antidote for CFS. In his article he recounted how he dissolved about five grams of licorice powder in milk and then drank the concoction. Within two hours he felt virtually cured.

Other researchers picked up on Baschetti's idea and began to explore the possibility that some cases of CFS are caused by low blood pressure and may therefore be treatable with licorice. They saw results, but only in patients who had enlarged lymph nodes, which can be associated with CFS. If it does turn out that CFS is somehow associated with adrenal insufficiency — in other words, low levels of aldosterone — then researchers can devise treatments based on hormonal supplementation. These treatments will be more reliable than licorice experimentation. But for now it goes without saying that anyone who wants to try licorice should do so only under a physician's supervision, because blood pressure and potassium levels must be monitored.

Prompted by the current interest in licorice, researchers at the University of Padua in Italy recently asked young men to eat seven grams of licorice tablets (0.5 grams of glycyrrhizic acid) a day for a week so that they could study the effects. The subjects' testosterone levels dropped by forty-four percent in just four days. Ouch! It seems that men who indulge in too much licorice may not be able to indulge their sweethearts. But single large doses may not have this effect. At least, not if we judge by Chaplin's shoe-eating episode. The actor married four times in his life, and all of his brides were under twenty years of age. No lack of testosterone there.

Chocolate Lovers Rejoice

On Wednesday, June 21, 2000 I achieved a personal best. I surpassed my own record for eating chocolate in a twenty-four-hour period. By a lot. I sampled chocolate truffles, praline tortes, molten chocolate cakes, double chocolate puddings, chocolate croissants, chocolate bars with labels that read like those on bottles of fine wine, and, to top it off, an unbelievable "secret chocolate marquise." I wasn't trying to commit suicide by chocolate. Quite the opposite, actually. I was upping my antioxidant intake. I had been fortunate enough to receive an invitation to speak at a symposium on chocolate organized by the American Chemical Society. The society had lured about two dozen food writers and editors from across the United States to the spectacular Belmont Estate outside Washington D.C. with an agenda that promised the latest updates on the science of chocolate as well as some appropriate gustatory delights. I don't think any of the participants were disappointed in either area.

I have always liked chocolate. I grew up on it. Every morning my mother blended a heaping spoonful of cocoa powder with a little milk, some sugar, and a touch of salt (to decrease the bitterness). Then she stirred the mix into gently boiling milk. This was my breakfast beverage. I was told that it was "good for me." I never questioned that assertion. Why should I? It tasted great. I really didn't see the need to put up a fuss as I might have when it came to, say, spinach. As I got older my love of hot chocolate paved the way towards chocolate bars — mostly Lindt and Suchard — but eventually my chemical education got in the way of these simple pleasures. I discovered that there was fat galore in a chocolate bar, fifty percent by weight, and lots of sugar, to boot. Not great for cholesterol or glucose levels. So I resigned myself to talking about chocolate instead of eating it. Describing the fascinating chemistry of chocolate turned out to be a great way of interesting students and the public in topics such as fermentation, food cravings, nutrition, and even brain function. And everyone, it seems, is captivated by the notion that chocolate contains a chemical, phenylethylamine, which helps us to fall in love. But, unfortunately, this just isn't so. The only thing that chocolate helps us fall in love with is chocolate.

Why is that? Why is chocolate the most frequently mentioned food in surveys about cravings? (Actually, this is mostly true for young women; men crave pizza.) While some scientists have argued that the cause is certain compounds, such as anandamide or caffeine, which do have the potential for pharmacological activity, the consensus is that chocolate is addictive because of its flavor. Not its taste, however. Flavor is more than just taste; smell and texture also come into play. Flavor chemist Dr. Sara Kisch proved this very effectively at the chocolate symposium. She handed out jelly beans of various flavors and asked us to sample them while holding our noses. They were indistinguishable. But identification was no problem once our nostrils were liberated. It was a dramatic demonstration of the role smells play in flavor detection. When the nose is pinched, no air can flow from the mouth to the nasal passage, where our smell receptors are located, and therefore the volatile compounds that are so vital to the production of flavor cannot be detected. That, incidentally, is why cold sufferers complain that their food doesn't taste right. Congestion has impaired their sense of smell.

The texture of chocolate is also important. Cocoa fat's melting point is just around body temperature, so solid chocolate rapidly turns into a smooth, luxurious liquid in the mouth. Some have argued that this constitutes a sign from God that we should be eating chocolate. I don't know about divine messages, but recently there have been signs from the scientific community indicating that moderate chocolate consumption may be good for us. Andrew Waterhouse of the University of California got the ball rolling in 1996 with his finding that cocoa beans are an excellent source of a class of chemicals known as polyphenols. These compounds had already received a great deal of attention because they'd been linked to health benefits and are prevalent in fruits, vegetables, and even red wine. Acting as antioxidants, they neutralize those nasty free radicals in our bodies. Free radicals are a by-product of breathing oxygen, and they have been implicated in diseases ranging from heart disease to cancer. People were thrilled that free radical fighters had now been found in chocolate, in significant amounts. Waterhouse even showed, in the laboratory, that cocoa extracts can prevent the oxidation of LDL cholesterol (the so-called bad cholesterol), a reaction that many consider to be a critical step in the formation of deposits in coronary arteries.

Then, in 1998, Harvard researchers dipped into chocolate and published a study that made headlines around the world. They tracked the health status of nearly eight thousand male Harvard graduates, all over the age of sixty-five, and concluded that those who ate one to three chocolate bars a month lived, on average, a year longer. More chocolate-eating, unfortunately, was not better. Men who ate one to three bars a week still did better than abstainers, but their life expectancy didn't increase by as much as the moderate chocolate consumers. Still, there was an effect. And many headline writers discovered that "Eat Chocolate and Live Longer" has a nice ring to it.

Dr. Joe Vinson of the University of Scranton believes that there is something to the chocolate effect, and he came to Belmont to tell us about his intriguing research. Vinson has determined the total polyphenol content of various chocolates and has also found a way of measuring how effective these mixtures are in preventing the oxidation of human LDL in a test tube. In other words, he has calculated a "phenol antioxidant index," which takes into account both the quantity and the quality of these desirable substances. At the symposium Vinson reported that cocoa powder and dark chocolate are the best, followed by milk chocolate. Instant cocoa mixes trail the field. Then Vinson delivered the kicker: chocolate has more, and better, polyphenols than fruits or vegetables; and more than red wine. A forty-gram bar of dark chocolate has as many polyphenols as a cup of that widely promoted antioxidant cocktail we call tea. But there is still the matter of chocolate's fat content. Researchers tell us, though, that at least half of it is stearic acid, which does not raise blood cholesterol.

Now you can see why we symposium participants were primed to partake of the chocolate delights served up by Belmont's superb chef. He even managed to make Shirley Corriher's secret chocolate marquise. Corriher, a biochemist who delighted us with her cooking-with-chocolate demonstration, had attempted to reproduce a marquise she had savored in a restaurant in Mionnay, France. The restaurant owner had been unwilling to divulge the recipe, but with her scientific background Corriher was able to figure out the nuances. She served us a square of smooth dark chocolate in a pool of sweetened cream. To die for. Maybe literally. I don't think the polyphenols were a match for the cream, butter, and egg yolks.

Although I can't recommend consuming treats like that marquise on a regular basis, I can at least say that there is no need to feel guilty about eating a few chocolate bars a month. Real chocolate — the darker the better. It may even make us live a year longer. More time to eat chocolate. As for me, I'm going back to my childhood. Hot chocolate made with defatted cocoa powder, here I come! Of course, I'm not forsaking fruits and vegetables, even spinach. But I must admit that I've wondered what would happen if I dipped a tea bag in hot chocolate for a polyphenol boost. Now, there's an idea for the next chocolate symposium. I hope they invite me.

Berry Good News

You don't worry too much about aging until you realize that you're doing it. And that realization usually comes when you first discover some stranger staring back at you from the mirror. The graying hair, the wrinkles, the hint of a double chin — it can't be you. What's going on? Inside, you feel the same as you've always felt, but now someone's trying to stuff you into the wrong body. Then you realize that time has flown, that things have changed, that you must finally abandon all hope of playing shortstop for the Yankees or scoring a goal for the Montreal Canadiens. Then the did-you-hear-about stories start. Heart attacks, tumors, arthritis, and diseases you never knew existed become the focus of your conversations. And your first question is always about the age of the wretched victim. Much too often that number hits close to home. You're walking a tightrope, you figure, and it's just a matter of time before you, yourself, become the subject of a did-you-hear-about. Better start practicing your tightrope walking. Just like those rats at the Center on Aging at Tufts University in Boston.

All right, so the rats weren't exactly walking a tightrope. They were balancing on narrow rods. Close enough. And they were doing it better after researchers fed them blueberry extract. To judge the age of a rat you measure the time it takes for it to lose its balance while walking the plank, as it were. By the age of nineteen months, which is roughly equivalent to the human age of sixty-five to seventy, a rat's balance time drops from thirteen seconds to five seconds. The creature also becomes less adept at learning to negotiate mazes, a real problem for a lab rat, since maze solving is a major part of its job description. All of this is understandable, since loss of coordination and a decline in short-term memory are part and parcel of the aging process. But hang on a minute! Well, maybe not a minute, but a few more seconds. Old rats managed to do just that — they increased their time on the rod from five seconds to eleven seconds after dining on blueberry extract for eight weeks.

Why blueberry extract? Because in previous laboratory studies researchers had tested forty major types of fruit and determined that blueberries possessed the highest antioxidant capacity. This means that the berries are particularly adept at neutralizing those rogue chemical species we call free radicals, those by-products of human metabolism we have linked with virtually every aspect of aging, from eye problems to brain deterioration. So it wasn't all that surprising that the rats' coordination and memory improved. In an effort to understand what was going on, researchers compared rats on the blueberry diet with rats on regular chow after exposing them all to high levels of oxygen for forty-eight hours. This is a standard procedure for the study of antioxidant effects, because breathing oxygen triggers free radical formation. A rat exposed to this much oxygen sustains roughly the same amount of free radical damage a human will sustain over seventy-five years.

The researchers' key finding was that the brain cells of the group on regular chow became less responsive to neurotransmitters, the chemicals that play a critical role in the functioning of the nervous system. Blueberry-fed rats did not show this impairment, in spite of the oxygen onslaught. Anthocyanins, the compounds responsible for the fruit's blue color, turned out to be the actual free radical scavengers. That's why the bilberry, a relative of the blueberry, may be even more effective — its flesh is blue right through, unlike many blueberry varieties, whose anthocyanins are concentrated in the skin. All of this rat research was enough to convince the study's lead author to take a capsule of bilberry extract daily and to pop blueberries like candy.

But my discussion of rat acrobatics isn't intended to turn people into blueberry-guzzling fiends. Eating blueberries won't save you dietary sinners. What matters most is the overall diet composition. In the rat study strawberry and spinach extract also produced improvements in memory, although those foods didn't have the same effect on balance and coordination as blueberries did. Still, the message is that fruits and vegetables have antiaging properties. As we are now learning, they harbor all kinds of useful chemicals. Like the B vitamin commonly known as folic acid.

Folic acid deficiency can cause mental retardation in babies. That's a fact. So researchers at the University of Kentucky wondered whether a deficiency later in life could be associated with the mental impairment that sometimes characterizes aging. They elected to study a group of elderly Minnesota nuns who were keen on contributing to the advancement of science. Nuns are good subjects for studies on aging, because they tend to have similar lifestyles and abide by the study's protocol. All these nuns had to do was take a daily folic acid supplement of four hundred micrograms. The results were intriguing. The nuns who were treated with folic acid were less likely to develop Alzheimer's disease, and autopsies of subjects who died turned up fewer brain anomalies among the supplement takers. There was another interesting finding, as well. The nuns provided researchers with samples of their writing going back to their youths, and these revealed that those who wrote well and expressed their ideas clearly had a lower risk of developing Alzheimer's. Use it or lose it, would seem to be the message. Rat studies back this up. When older rats are given toys to play with, their brains manifest physical changes that are typical of younger animals.

Consuming lots of fruits and vegetables and exercising our brains will, beyond a doubt, prove beneficial to us, but isn't there something more dramatic we can do to slow the aging process? Maybe one day. Keep the term "telomerase" in mind. This enzyme encourages cells to keep dividing indefinitely, instead of the usual seventy-five or so times. We may eventually find it possible to insert into a cell a gene that codes for the production of this enzyme. Sounds tantalizing, but there would be problems involved. What if the birth rate greatly exceeded the death rate? The planet's resources and economy could be ruined, its environment ravaged. We'd better be careful what we wish for, because we may get it.

Still, I'm eating my spinach and my blueberries. And how about that study from the University of California indicating that scientists age more slowly if they really like their jobs? That's good. But the study also showed that unsociable scientists live even longer. Hmmmm. So don't invite me to any parties. Not even if you're serving spinach and blueberries.

Just Give Me the Flax

There is a bagel bakery in Chicago that sells "brainy bagels." I kid you not. They're made with flaxseed, which has an extremely high content of alpha-linolenic acid, a compound that falls into the category of omega-3 fatty acids. There is some evidence that omega-3 fats (the terminology refers to an aspect of their molecular structure) may be linked to mental acuity, hence the creation of the brainy bagels. Although I may be skeptical about the bagel's ability to oil our intellectual machinery, I think it's probably a smart thing to eat flaxseed bagels, flaxseed muffins, or flaxseed anything. I know what you're thinking. We already have to eat oat bran, soybeans, and broccoli for good health. Now flaxseed, too? I'm afraid so. The plant that gave us linen, linseed oil, and linoleum may also give us longevity. You've got to be impressed when scientific studies show that flax may aid in the prevention or treatment of many conditions, including heart disease, cancer, and diabetes. How could a simple seed do all that? Because the chemical makeup of the seed is complex; it has numerous components, but three, in particular, look promising to health researchers.

Aside from its murky role in brain function, alpha-linolenic acid has some well-documented effects on heart disease. It reduces the risk of blood-clot formation, lowers the chance of potentially lethal irregular heartbeat, and probably has an anti-inflammatory effect on blood vessels. Several studies have found a link between increased intake of the acid and reduced risk of death from heart disease. Alpha-linolenic acid helps lower blood cholesterol, but flaxseed has another component that performs this function even more effectively. This is soluble fiber, which isn't absorbed by the body and passes right through the digestive system. On its journey through the digestive tract, the fiber binds cholesterol and prevents it from being absorbed. Furthermore, it also binds the bile acids needed for digestion, forcing the body to produce more. Since the starting material for the biosynthesis of bile acids is cholesterol, blood levels of cholesterol go down.

So how much flaxseed do you have to torture yourself with in order to reduce your cholesterol? About two heaping tablespoons (twenty to twenty-five grams) a day should do it. Grind the flaxseed in a coffee grinder, otherwise much of it will come out in the same form as it went in. You can store the ground seeds in a tightly closed jar in the fridge, but you can't keep it for very long. The flaxseed oil turns rancid pretty quickly. How do you eat the stuff? Sprinkle it on cereal, mix it into yogurt, stir it into juice. The total cholesterol-lowering effect is usually in the range of five to ten percent, with LDL, the "bad" cholesterol, dropping by as much as eighteen percent. This is close to the results we can obtain with medications. And there's another benefit. Flax, which has insoluble fiber in addition to soluble fiber, can eliminate more than just heart disease. Those who ingest it will find themselves visiting the facilities regularly. They'll have no need for prune juice. And the fiber is even effective in regulating blood sugar in diabetics.

Now we come to the most exciting compound in flaxseeds — namely, secoisolariciresinol diglycoside (SD). Don't let the name scare you, because this is potentially great stuff. Bacteria in our colon feed on it and then churn out compounds known as lignans. These have a chemical similarity to estrogen, and, since they originate from a plant source, they're termed "phytoestrogens," from the Greek *phyto* for "plant." Why all the excitement? Because these phytoestrogens have anticancer properties. Our attention was first drawn to this fact when a study conducted in Finland showed that the urine of women with breast cancer had lower levels of lignans than that of healthy women. By contract, the study found that the highest levels occurred in women living in areas with low breast cancer rates. This makes sense if we look at the molecular mechanism of estrogen-responsive breast cancer. In this common form of the disease, circulating estrogen interacts with special proteins in cells called estrogen receptors. The long protein molecules are coiled into a specific shape, with inherent cavities into which estrogen molecules fit the way a key fits into a lock. As the end result of this interaction, the genes in the cells that are responsible for cell proliferation are activated. In other words, the cells start multiplying rapidly, a situation that can lead to cancer. The more estrogen molecules there are, the more cells are activated.

Phytoestrogens are chemically similar enough to estrogen to fit into the receptors, but the fit is not perfect — something like a rusty key being inserted into a lock. It may not open the lock, but it certainly prevents the right key from entering. If some of the locks are occupied by these rusty keys, there is a smaller chance of cell proliferation. But does this fanciful analogy have any practical significance? That's what Dr. Lilian Thompson of the University of Toronto wanted to find out, so she induced cancer in rats with a known carcinogen and fed some of the rodents varying amounts of flaxseed. Fewer and smaller tumors developed in the flax-fed animals. Even when tumors did form, they progressed more slowly and invaded tissue less vigorously. This finding meshed with the previous observation that feeding flax to female rats lengthened their estrous cycles. Estrogen governs the cycle, and if there is less estrogenic stimulation, then the cycle is longer. With less stimulation, one would expect a reduced risk of cancer, as Thompson indeed found.

But more flax is not necessarily better. When the rat chow contained five percent flax by weight, there was delayed onset of puberty, but when the amount of flax was doubled onset of puberty came earlier. In males, five percent flax reduced prostate cell proliferation, but ten percent increased it (but men would never assume a diet containing 10% flax by weight). How do we explain this? It's actually quite straightforward. Although rusty keys may not open locks as effectively, if you have enough of them, they will start opening some locks. Obviously, with flax, as with other physiological interventions, the dose is important. But rats are rats, and they are different from humans. What human data do we have? There is no question that eating flax affects body chemistry. The lignan content of urine rises with increased flax consumption, so lignans are certainly making it into the bloodstream. Not only that, flax also changes the way that the body metabolizes natural estrogen. In a University of Minnesota study, researchers put postmenopausal women on diets supplemented with flax, and the estrogen metabolites in their urine were measured. The results showed a significant increase in the level of an estrogen metabolite that protects against breast cancer.

Studies with mice have even shown effects against other cancers, like melanoma. When researchers induced melanoma in the mice, the tumors they developed reduced when their flaxseed dose was increased. Perhaps even more interesting, the tumors did not metastasize effectively, suggesting that flax may be a good nutritional supplement for cancer patients.

Yet what we really need is a well-controlled study of human cancer patients who supplement their diet with flaxseed. Just such a study is under way at the University of Toronto. One hundred women with newly diagnosed breast cancer have been enrolled, and each day half of them eat a muffin containing twenty-five grams of flaxseed. Results so far show a decrease in breast-cancer cell proliferation. But do we really have to wait for the final outcome of such a study to increase our flax intake? There's no reason we should. We've seen enough evidence concerning the benefits of fiber and improvements in blood sugar and cholesterol counts to warrant flaxing our diets to the extent of about two heaping tablespoons of ground seed a day. I'm not sure how much flaxseed the brainy bagels of Chicago have, but it may be a fair bit. When a TV news reporter asked a young bagel shop patron if he felt any different after eating one, he replied, "Yup, I feel smarter." Maybe he really was — he then proceeded to eat another flaxseed bagel.

Remember This

One of the most worrisome aspects of aging is memory loss. We can put up with wrinkles, sagging skin, aches and pains, but the first time we forget a name it conjures up horrific visions of Alzheimer's disease. Most people, of course, do not go on to develop this dreaded disease, but they may very well have to put up with some memory loss as a consequence of senior citizenship. Can we do anything about it? Is there some pill we can pop to improve memory?

These days everybody is talking about ginkgo biloba, a concentrate made from the extracts of the leaves of a tree that has been around for over two hundred million years. The species was brought from China to Europe in the eighteenth century, and it became a popular decorative tree. It also serves a medicinal purpose. The leaves are picked, dried, and ground to a powder. This powder is extracted with an acetone-water mixture, and then these solvents are evaporated off. The residue that remains is compacted into a variety of pills. Several components of this residue may have beneficial effects; in particular, they may improve circulation to the brain. Clinical studies have confirmed that ginkgo extracts dilate blood vessels and improve blood flow in arteries and capillaries. Two classes of compounds, namely the flavonoids and the terpenes, are of interest in this regard. But, as one would expect, the ginkgo residue incorporates numerous compounds, and its composition can vary depending on how it was processed. So, if ginkgo is to be used medicinally, then it must be standardized. In other words, we need to know what and how much of the active ingredients we are taking.

Modern chemistry allows us to carry out such standardization, and standardized pills are available; in fact, they have long been available in Germany, where ginkgo is a top-selling drug. Manufacturers label the standard pill as containing twenty-four percent flavonoids and six percent terpenes. Most studies that have examined the effect of ginkgo on memory used this formulation. So much for the background. What we are interested in is what these studies show. First the good news: a number of studies have shown that ginkgo extracts improve memory. Then the bad news: the improvement has been modest, to say the least.

Probably the best study carried out so far involved 309 patients who either had Alzheimer's disease or had suffered a stroke that affected their memory. They were randomly divided into two groups. One group ingested 120 milligrams of ginkgo, the other a placebo. About one-third of the patients taking ginkgo showed some improvement on some performance tests — hardly a result to shout about. The best results occurred among the least impaired patients, suggesting that perhaps ginkgo plays a role in postponing dementia. Some proponents of herbs believe that ginkgo may improve tinnitus (ringing in the ear) as well as vertigo although controlled studies have failed to demonstrate this.

Is there any risk in trying this supplement? It would appear to be minimal. Although ginkgo may cause such side effects as mild gastrointestinal problems, headaches, and allergic reactions, the fact is that these effects showed up to the same extent in the placebo groups involved in the ginkgo studies. The only worrisome aspect of ginkgo is that it inhibits blood clotting. This would explain three serious cases of internal bleeding reported in the scientific literature. Of course, three cases out of millions of patients is a practically irrelevant percentage — unless you happen to be one of these cases yourself. It also means that ginkgo can represent a risk for anyone who is already taking medication that thins the blood, such as aspirin or coumadin. Such people should consult a physician well versed in the use of ginkgo before taking the supplement.

The studies I refer to here have dealt with patients already suffering serious memory loss due to disease. But what about ginkgo for reducing normal age-related memory loss? If you watch TV ads for the stuff, you'll be convinced that gingko is just the thing to improve virtually anyone's memory — homemakers maintain that they can remember their shopping lists better; students imply that they perform better on exams. Yet there is precious little evidence to back up these claims. In one study eight women who took a huge dose (six hundred milligrams) of gingko experienced some memory improvement, but this could not be duplicated. In another typical study, after taking the usual dose of 120 milligrams a day for three months, twenty-seven people performed better on one out of thirteen memory tests than twenty-seven others who took a placebo. Not an earth-shattering result.

So the bottom line on ginkgo is not very encouraging. While some tests have shown that it can lead to modest memory improvement in demented patients, the effects are far from spectacular. Anyone wishing to give ginkgo a try should consider taking only the standardized versions. Check the label for something like twenty-four-percent potency, and remember to take the pills daily, because it will probably take a month for the results to manifest themselves.

Substances do exist that haven't been talked about as much as ginkgo but that hold greater promise of alleviating memory loss. Phosphatidylserine, for one, which resides in the membranes of all nerve cells and seems to play an important role in transmitting information. In a study of 149 normal volunteers aged fifty to seventy-five, half were given three hundred milligrams of phosphatidylserine for twelve weeks, and half got a placebo. Those taking the phosphatidylserine showed significant improvement in learning and recalling names, faces, and numbers. Perhaps more attractive than taking pills is a Japanese proposition that sake can enhance memory. Researchers have discovered that something in the rice wine inhibits an enzyme in the brain that has been linked with impaired memory. I don't know about this one. I had several friends in university who used to be into the sake, and I remember that they couldn't remember their sake binges very well the morning after.

Speaking of university students, several studies have shown that odor can stimulate the memories of people writing exams. Researchers asked students at Yale University to write down the opposites of forty common adjectives while imagining the smell of chocolate. They were not told that the next day they would be required to remember the associations. The researchers then repeated the experiment, this time exposing the students to the smell of chocolate during the word exercise, during the recall, during both processes, or during neither. Those exposed during the exercise and the recall did significantly better. The chocolate smell is pleasant, but it isn't necessary; even mothballs will work, as long as the same odor is present during the exercise and the recall phases. Can students benefit from using a different smell for each topic? Apple for physics and peach for chemistry? What about training pilots for air emergencies using odor association? The mind reels with the possibilities.

Another important element of memory improvement is exercise. That's right — the body requires exercise and so does the brain. Crossword puzzles, bridge, and chess are great. So is reading stimulating scientific information. Which I hope this is. Trying to remember what you've just read will provide you with a good dose of mental gymnastics. And now I think I've told you almost everything I remember about memory loss. I think, though, that I used to know more. Maybe I need some ginkgo pills. But I don't think I could remember to take them.

Cheers for Sour Cherries

When I was young I didn't consider the first official day of summer to be June 21st. It didn't matter what kind of weather we were having, summer did not really begin until the day my mother made cold sour-cherry soup from the first freshly picked tart cherries of the season. What a delight it was! I wolfed down bowl after bowl because it tasted so good; I never dreamed that it might actually be good for me.

Okay, okay. I can see your eyebrows rising out there. Here comes some more unwelcome dietary advice about what we should be eating. You're probably sick of it. Depending on which particular research findings have made their way into the news in a given week, you've been urged to consume tomatoes, broccoli sprouts, Brazil nuts, margarine, soy milk, apple-cider vinegar, oats, flaxseed, emu meat, kiwi, red wine, cheese, tofu, or some herbal extract that nobody has ever heard of. The advice is usually accompanied by reference to some "breakthrough" research by a "leading" scientist, and it will trigger a whole new feeding frenzy, at least for a few days. Then another equally eminent spoilsport authority will blast holes in the research and explain that the findings may apply to rats raised on laboratory chow or to humans who subsist on baloney sandwiches but cannot be extrapolated to the public at large. So everyone goes back to eating exactly what they were eating before.

Too bad, because buried in that scientific and pseudoscientific cacophony are some nutritional gems. They just have to be scooped out by those willing to scrutinize the underlying science. The benefits of soy and broccoli, for example, are well established, but these aren't miracle foods. We have to evaluate them in the context of the overall diet. So, I'm certainly not going to get into the good-food/bad-food game and suggest that people load up on certain foods and eliminate others. As the saying goes, there are no good foods or bad foods; there are only good diets and bad diets.

That being said, we have to acknowledge that diets are composed of individual foods and that some foods should play a more prominent role than others. Perhaps the best-supported nutritional advice we can follow is to consume ample quantities of fruits and vegetables. That's because they are packed with various phytochemicals (chemicals derived from plants), which numerous studies have linked with disease prevention. Scientists estimate that we could reduce the incidence of cancer by about twenty percent by simply consuming five to ten servings of fruits and vegetables daily. Of course, we would love to know which specific chemicals might have this effect and where they are most likely to be found. Then we could choose more wisely while food shopping and eating and perhaps even incorporate certain phytochemicals into our dietary supplements.

Now, back to my cherries. These fruits have been studied extensively, particularly the tart varieties. Much of the work has been carried out by Dr. Muralee Nair and his group of researchers at Michigan State University, undoubtedly prompted by the fact that the State of Michigan produces about seventy-five percent of the U.S. crop. Obviously, any positive findings about cherry eating are most welcome in Michigan, and there certainly have been some. The focus of fruit research these days is on compounds referred to as antioxidants. Many researchers believe that a major proportion of the benefits attributed to consuming fruits and vegetables can be traced to the presence of antioxidants, which can neutralize the effects of free radicals. These highly reactive substances, which we have already encountered on several occasions, are a by-product of normal metabolism, and researchers have linked them with conditions ranging from cancer to arthritis to ageing. They also play a role in the spoilage of foods, particularly those with a high polyunsaturated fat content.

Anthocyanins, the pigments found in many plant products, are very effective antioxidants. Indeed, when isolated, their ability to neutralize free radicals is comparable to that of BHT, a widely used food preservative. This observation prompted the Michigan State team to investigate whether the addition of sour cherries to hamburger meat could reduce the rate at which fat was oxidized. Fat oxidation in foods may have adverse health consequences, and it is a major cause of flavor deterioration. The cherries did the job. In fact, they did more. One of our concerns about heating meat is that it causes the formation of heterocyclic aromatic amines (HAAS), compounds that are carcinogenic in rodents and primates and probably in humans. When the researchers mixed cherries into the hamburger to the extent of about twelve percent by weight, they managed to reduce the formation of HAAS by a staggering seventy to eighty percent. Whether this was due to the anthocyanins or to some other compounds that occur naturally in cherries, they have yet to ascertain.

A Michigan cherry grower who also happens to be a

butcher (how often do you find that combination?) has already capitalized on the idea. He convinced the U.S. Department of Agriculture's National School Lunch Program to feature cherry burgers, and these delicacies already grace school menus in sixteen states. They are low in fat and high in antioxidants — and the word is that they taste great, to boot.

Research on the antioxidant properties of sour cherries had another salubrious outcome: the isolated antioxidants, particularly one called cyanidin, unexpectedly proved to have an anti-inflammatory effect. Inflammation is generally caused by compounds in the prostaglandin family, which are formed in the body through the work of enzymes called cyclooxygenases. Aspirin, for one, inhibits these enzymes, and the widely used anti-inflammatory is therefore referred to as a prostaglandin inhibitor. It seems that cherry extract is an even more potent prostaglandin inhibitor, perhaps by as much as a factor of ten. In a preliminary study, researchers provided arthritis sufferers with twenty sour cherries as an alternative to a standard dose of aspirin. Sour cherries are seasonal and not consistently available, but sour-cherry juice isn't so hard to find. Whether the juice is as effective as the fruit itself has yet to be determined, but, hey — it sure tastes good, so what do we have to lose? I haven't yet seen any cherry-extract pills in the health food stores, but I'm sure some enterprising merchants are already working on putting the cart before the horse.

If you would like a little more tantalizing cherry information, then here it is. Quite aside from the antioxidants, cherries contain another substance with interesting properties. Research has shown that perillyl alcohol produces antitumor effects in laboratory animals and that it has cancer-prevention properties. In preliminary human trials researchers have not yet seen a significant effect in terms of treatment, but perillyl alcohol does hold promise as a model for a new class of antitumor agents.

So, life may really be a bowl of cherries. And the sour cherries may have the sweetest effect. I'm not advocating these fruits as a cure-all, but why not add them to our growing arsenal of fruits that have disease-fighting properties? Just don't eat the seeds — they contain cyanide. Certainly not enough to do any harm, but spit them out, anyway. As far as you can. With a little practice you may even have a chance of winning the International Cherry Pit Spit Competition, held every year in Eau Claire, Michigan. Actually, bringing home the trophy may require more than a little practice. It may demand some elaborate body, neck, and mouth coordination. According to *The Guinness Book of World Records*, the distance to beat is twenty-two meters. It may take some perseverance, but after eating all those cherries you won't have to worry about muscle inflammation.

Bitter Water and Sweet Science

In 1618 an English farmer noticed that he could lead his cows to water but could not make them drink. He tasted the water, which had been drawn from a well, and he realized that there was some wisdom to the cows' behavior. The water tasted terribly bitter. It occurred to him that something that tasted so awful must be good for something, and he was right. He used the water for a hot, soothing foot bath and later noticed that it healed scratches and skin rashes. The science-minded farmer then discovered the liquid's most dramatic effect. He drank a whole glass of it, and — let's just say that the result was a rather hasty trip to the outhouse. What was so special about this water? It didn't take the farmer long to figure it out. After he'd allowed the water to evaporate he was left with copious amounts of white powdery residue. "Epsom salt," he called it, and the name stuck. That well at Epsom eventually became a huge commercial success.

We still rely on Epsom salt to soothe inflammation by withdrawing water from muscles and tissues; some use it as a laxative, as well. These days we know what the active ingredient is: magnesium sulfate. But the English farmer wasn't the first to notice the bitter taste of the substance. That honor may well go to Moses. The Bible tells us that, after leaving Egypt, the Israelites wandered about the desert with very little food or water. One day, much to their delight, they came upon the pond of Marah. Their happiness was short-lived, because the water turned out to be so bitter that they couldn't drink it. Moses then turned to his research director and was rewarded with some good advice: "The Lord showed him a tree; and when he cast it into the waters, the waters were made sweet." What happened here? The main components of wood are cellulose and lignin. When exposed to the sun, these compounds develop ion-exchange properties, meaning they gain the ability to absorb both magnesium and sulfate ions. So, Moses ended up purifying the water, making it "sweet" to drink.

The Israelites' chemical escapade may not be the only instance in which magnesium is associated with divine intervention. In fact, there would be no life on earth without magnesium. It is an integral part of the chlorophyll molecule, which is responsible for photosynthesis. No magnesium, no chlorophyll, no photosynthesis, no plants, no life. And what eventually happens to the magnesium in chlorophyll? It depends. Sometimes you'll notice that when you cook green vegetables they lose a bit of their color. This is especially true of peas, which often end up looking rather anemic. The cooking process releases natural acids from plants, and these tend to remove the magnesium from chlorophyll, destroying its color. The time-honored remedy for this is to add a little baking soda to the cooking water. The soda neutralizes the liberated acids and maintains the bright color. For more on this see "Green Peas, Please."

We may take that color as a good sign — it means that the magnesium hasn't leached out into the cooking water. In truth, though, in the context of our overall diet, the amount of magnesium we miss out on by eating pale peas is insignificant. So, why should we be concerned about missing out on magnesium at all? Because it is an essential mineral. It is critical for the functioning of our nervous system, our heart, the regulation of our blood pressure, and even bone formation. Magnesium deficiency can cause leg cramps, weakness, vertigo, twitching, irritability, and even irregular heart beat. Indeed, studies have linked the drinking of soft water — which is low in minerals like magnesium and calcium — with an increased risk of heart disease.

Luckily, it isn't hard to get a good dose of dietary magnesium either directly, by eating plants, or indirectly, by eating animals that have dined on plants. Whole grains, seeds, nuts, and beans, particularly soya, are excellent sources of magnesium. Even chocolate has some. In cases of magnesium deficiency, or sometimes in an attempt to treat persistent fatigue or irregular heart beat, physicians will suggest magnesium supplements. Magnesium gluconate is the least likely to cause diarrhea, but even with this supplement we must take care. Daily doses of two hundred milligrams of magnesium are safe enough, but high doses can actually cause heart problems. Termite droppings, of all things, happen to be very high in magnesium, and it is interesting to note that since ancient times certain African tribes have treated fatigue with little balls of the stuff. I think I'd rather take Epsom salts. Or even eat tofu.

So far I have referred exclusively to compounds of magnesium, that is, to substances in which the magnesium is chemically combined with other elements. The pure metal itself was first isolated in 1808 by the brilliant English chemist Humphrey Davy. His source was Magnesian stone, a mineral that had been mined in the Magnesia district of Greece since antiquity. Today we know it better as soapstone. Magnesium turned out to be a very useful metal because it was strong and light. It was also flammable and burned with a brilliant flame. Early photographers put this property to good use by developing flash powder, which, when ignited, generated enough light to allow them to take pictures indoors. Later, during World War II, magnesium was used in the manufacture of incendiary bombs. Modern technology has found ways to alloy magnesium with other metals to reduce flammability while maintaining strength. The metal's low density makes it ideal for airplane fuselage and rocket-casing construction; it's also the stuff they use to make mag wheels for cars.

Magnesium wheels are a real status symbol, but only if they're clean. Car buffs can choose from a variety of available cleaning agents, but one wheel cleaner, formulated with ammonium bifluoride, has to be handled with extreme caution. The chemistry is interesting, and so are its consequences. Ammonium bifluoride is not the active agent — it breaks down to form hydrofluoric acid and fluoride ions. While this mix produces an effective cleaner, it can be quite dangerous. The product is marketed as a spray and can be accidentally inhaled. The fluoride reacts with calcium and magnesium in the body, causing a drop in blood levels of these minerals, and possibly leading to cardiac arrest. In a few cases children who played around with the spray ended up in hospital with magnesium deficiency.

We've come a long way towards understanding magnesium since Moses cast that piece of wood into the water, and the magnesium story continues to evolve. Researchers are studying the effects of administering magnesium sulphate intravenously during a heart attack to reduce the risk of arrhythmia. Some studies have already shown a reduced incidence of mortality. No doubt our English farmer would be most surprised to discover that his bitter water is fulfilling such a sweet role.

Feeding the Diet Industry

Over 26,000 weight-loss diets have been published in this century. So, now, ask yourself this question. What are the chances that, let's say, number 26,571 will succeed where the 26,570 that preceded it have failed? If just one of those diets actually worked, then there would be no need for new diets, would there? Anyone wanting to lose weight could just adopt the successful scheme and that would be that.

We can't attribute our lack of a successful diet scheme to a general lack of effort. Eminent scientists and nutritional nobodies alike have had a go at the problem. After all, there is a lot of money to be made. In North America alone, overweight people spend close to forty billion dollars a year on girth control. Plenty of entrepreneurs want a piece of that pie, and so we are subjected to a steady diet of articles, books, and products claiming to have found the answer that has eluded everyone else.

We've seen everything. Low carbohydrate diets. Low fat diets. High protein diets. Diets based on the signs of the Zodiac (don't even ask). Diets dominated by cabbage soup or apple cider vinegar. Diets that permit only raw fruits before noon (they tell you not to worry about the diarrhea). Diets that prohibit mixing proteins with carbohydrates. Diets that count only grams of fat. Diets that count only calories. Diets that don't count calories at all. Diets that guarantee weight loss if you eat nothing but unprocessed alfalfa sprouts while hopping on one leg rhythmically to the strains of the theme from *Rocky*.

The result of all this dietary flimflam is that North Americans are fatter than ever. The flesh that bursts out of tank tops on a New Jersey beach or rolls out of shorts two sizes too small at Canada's Wonderland proves that we're losing the battle of the bulge.

How can this be? Just look at the labels on products that line the shelves of your local supermarket: low fat, zero fat, reduced fat, pseudofat. You would think we'd all be withering away. Yet as fat consumption decreases, the percentage of overweight people rises. Mysterious? Not really. Fat consumption may have gone down, but calorie intake has zoomed upwards. And, contrary to some of the popular rhetoric, calories *do* count.

Let's get somcthing straight. Calories are not things. They cannot be full or empty. They cannot be burned. A calorie is nothing more than a unit of measure. If you want to get technical, then a food calorie is the amount of heat needed to raise the temperature of one kilogram of water from 14.5°C to 15.5°C. So, then, what does it mean that a doughnut contains two hundred calories? Simply put, it means that if it burned completely in the presence of oxygen in a device called a calorimeter, it would generate enough heat to heat two hundred kilograms of water by one degree.

Our body, like any machine, requires a source of fuel. We talk, we breathe, we eat, and sometimes we even think. All of it requires energy, and this we get from burning food. Not in a furnace, but in our cells. As the food burns — that is, as it converts to carbon dioxide and water — it releases energy to maintain body temperature and to power all of our activities. If, on the one hand, we eat but aren't very active, the unused fuel piles up, and we put on weight. If, on the other hand, we don't take in enough fuel, the body switches to burning its stored supplies, and we lose weight. Pretty simple.

Designing a diet that causes weight loss is therefore easy. Each day, eat a variety of foods that together contain fewer calories than the body needs, and weight loss will ensue. That's why all diets are successful in the short term. The average adult human has a resting metabolic rate in the ballpark of 1,400 to 1,700 calories. That means that even when we are completely at rest, we require this many calories within a twenty-four-hour period just to stay alive. If we do not consume the adequate number of calories, the body is forced to draw upon its stored supplies. It doesn't matter where your calories come from. You could go on a chocolate diet and lose weight, as long as your intake was in the range of twelve hundred calories. But how long could one stay on a chocolate diet? (I know, I know — a long time.)

So losing weight is not difficult. But keeping it off is another matter. Within a short time the dieter's body will sense deprivation and employ the emergency measure of lowering the resting metabolic rate. Then the body will become more efficient and require fewer calories. Weight loss stops. At this point about ninety-five percent of dieters will increase their calorie intake, because a low calorie diet just doesn't deliver enough entertainment for the mouth or satisfaction for the stomach. Successful dieters are those who increase their level of exercise: this increases the resting metabolic rate, upping calorie expenditure. Dieters who don't do this will regain the weight they've lost, plus about ten percent interest. As their pants get tighter, their frustration builds up, and they will embark on yet another diet, only to achieve similar results. This is yo-yo dieting, and it's dangerous. Studies have shown that yo-yo dieters run a seventy percent higher risk of dying from heart disease than people whose weight stays steady, even if they are overweight.

Diet gurus, without admitting it — or often without realizing it — devote much energy to searching for ways to keep people who are coping with a reduced calorie intake from craving more food. The current hot idea is that one can do this by lowering insulin levels in the blood. These days a number of diet gurus are stepping on each other's toes, each claiming to have discovered the magic. We have the Montignac Diet, the Protein Power Diet, the Zone, and Sugar Busters — they're all the rage. Even noted celebrity scientist Suzanne Somers has entered the fray, advising us to "Somerize," much to the annoyance of Michel Montignac, who has taken legal action against her for infringing on what he regards as his own inspiration. All these diets share one feature: they outlaw foods that have a high glycemic index, causing a rapid rise in insulin, which, the authors claim, stimulates the appetite and prompts fat storage. Virtually no sugar is allowed. Even a bagel, because of its starch content, becomes a nutritional pariah. A low carbohydrate diet, not a low fat diet, is the key to salvation and weight loss, the gurus proclaim. This may be of interest to the one billion Chinese, virtually none of whom are overweight, in spite of the fact that they eat one third as much fat and twice as much starch as Americans.

Now, I think there is something to the insulin business, but it's not the Holy Grail. While cutting back on refined sugar and refined flour is a good idea for a number of reasons, does it guarantee weight loss? I don't think so. Dr. Robert Atkins has been pushing an extreme version of this diet, recommending high protein foods such as steak and eggs and bunless hamburgers (surely a delight), for twenty-five years. He claims to have 25,000 satisfied patients. Yet he hasn't published this data. Isn't that odd? If his claims were true, then he would have accomplished the nutritional equivalent of turning lead into gold. He'd merit a Nobel Prize.

So, are you disappointed? Fret not. There is an answer to the diet conundrum. Come, closer — let me whisper the secret to you: Eat less and exercise more! But some of you may not like that advice and would rather wait for diet number 26,572, which promises weight loss as long as you eat nothing but pickled frog legs and curried pigeon toes as you stand on your head and whistle into the wind.

Green Peas, Please

"Those peas look really yucky!" my little daughter whined as she pushed her plate away. She was right. They did look kind of unappealing. I was determined to get her to eat her vegetables, but how was I going to fight this seemingly unwinnable battle? The time had come for us to play with her food.

We took two identical pots and placed some water and a cupful of frozen peas in each. To one we added a little baking soda, and then we heated both pots to the boiling point. At first the peas in both pots turned greener, but it soon became evident that the peas with the added baking soda were retaining their luscious green color, while the others were fading to an unappetizing pale hue. "That's neat," my daughter said, still eyeing the peas suspiciously.

It really is neat, especially when we understand the underlying chemistry. These days it's certainly appropriate to discuss green chemistry; there are green parties, green coalitions, and green products. Our collective aspiration to greenhood has become a noble undertaking, an attempt to mitigate our failure to protect the natural environment and, consequently, our quality of life. And green really does mean life. Without chlorophyll, the green pigment in plants, there could be no photosynthesis, no peas, and no life as we know it.

Sunlight is composed of violet, indigo, blue, green, yellow, orange, and red rays. On occasion water droplets in the atmosphere will separate white light into individual bands of these colors and we'll see a rainbow. Other substances can either absorb or reflect the different colors of light, making our world very colorful. A red apple, for example, reflects red light and absorbs all other colors. So, as far as plants are concerned, the color green is unimportant. They don't need green light, and therefore they reflect it. The shade of green we see depends on just how the light is reflected. We may look at a freshly mown lawn and see stripes, because the mower bends the grass towards us and then away from us as it traverses the area; the blades of grass bent away reflect more light and therefore appear to be a lighter green.

What is important, as far as the existence of life on earth is concerned, is not the reflected light but the light that is absorbed by green plants. Chlorophyll absorbs this light and converts it into the energy it needs to drive the process of photosynthesis. Glucose, produced from carbon dioxide and water through photosynthesis, is the raw material from which virtually all other plant components are made. If there is no chlorophyll, there is no photosynthesis, there is no plant, and ultimately no life. The color green, therefore, provides an apt symbol for the environmental movement. But we have always known that green represents something special. We've searched for greener pastures, and we've been convinced that the grass is always greener on the other side of the fence. We're interested in Greenpeace, and of course, green peas.

So what was going on in those two pots of peas? When the peas were first deposited in the cooking water, tiny air pockets on their surfaces, which normally dim reflected light, collapsed as the air from them escaped. Then chemistry took over. Chlorophyll is a large, complex molecule that houses a magnesium atom at its center. Unfortunately, acids readily displace the magnesium, replacing it with a hydrogen ion and resulting in a much less colorful compound. Vegetables contain a number of naturally occurring acids, which are released during cooking and prepare to wreak havoc on chlorophyll.

A possible solution springs to the chemist's mind. Why not neutralize the acids with a base before they can react with chlorophyll? Cooks have actually been doing this for eons without really understanding the underlying chemistry. They toss in a pinch of sodium bicarbonate, or baking soda, to neutralize the acids and keep their veggies looking green. We can achieve a similar effect by cooking the vegetables in an uncovered pot, using a large amount of water. The acids then either volatilize into the air or dissolve into the cooking water to form an acidic solution so dilute that the chlorophyll molecules survive. As far as adding baking soda goes, you only need a little. If you add too much, your vegetables will turn mushy and some of the vitamins will be destroyed. "Good," my daughter declared upon hearing this; she was happy to take revenge against the vitamin pills she was regularly forced to swallow. Still, her interest hadn't waned, so the private lecture continued.

Next we took another sample of peas and boiled them, but this time, instead of adding baking soda, we plopped a penny into the pot. The green color did not fade. This ingenious technique was first introduced in a cookbook published in 1751. The acids cause a little copper to dissolve, which displaces magnesium from chlorophyll. This process alters the green color, but by replacing magnesium with copper we create an acid stable compound. In the old days people would drop pennies into pickle jars as a means of keeping their pickles looking fresh and green. I thought this was also pretty neat, but I could see that by now my daughter's fascination with the green was fading. We needed a more dramatic experiment.

Out from the freezer came the green frozen shrimp. Now these were *really* "yucky." But the little eyes opened wide when I plunged the shrimp into hot water — in seconds they turned a bright red. This was better than the pea performance. What was going on? Believe it or not, shrimps and lobster have something in common with carrots: all contain a good dose of carotenoids. Carotenoids are yellow-orange compounds that are quite widespread in nature, but the first one isolated was beta carotene from carrots, which then gave its name to the whole family. These compounds are responsible for coloring oranges, red peppers, watermelon, tomatoes, egg yolks, apricots, corn, pink grapefruit, pink salmon, and pink flamingos.

Lobster and shrimp dine on carotenoid-containing plankton, and the compounds become concentrated in their shells. Here the carotenoids are bound up with protein molecules, and the carotenoid-protein complex has a dark green color. When the protein is heated, it is denatured. In other words, it breaks down and disassociates from the reddish pigment, astaxanthin, which then becomes visible. To a smaller extent this is also evident in cooked carrots, which become more orange than they were before. This was another experiment my daughter and I decided to try. We cooked up some fresh carrots to see if they would become more orange. They did, but the effect was not as pronounced as it was with the shrimp, because carrots have little protein.

The moment for our final experiment had arrived. Would my daughter eat her peas, and maybe even a few carrots, after we'd made them scientifically exciting? Nope. At least not until my wife piped up, "Will you eat the peas for Mommy?" Down the hatch. I flushed red then turned green with jealousy.

CHEMISTRY HERE, THERE, EVERYWHERE

The Genie in the Bottle

I have a particular interest in genies. It isn't often that you encounter one, so imagine how thrilled I was when I did. All right, so she wasn't a real genie, but she did play one on television. It happened at a trade show in Vancouver, where I was appearing on behalf of the Discovery Channel. Right next to us was the Arts and Entertainment booth, and that network's featured guest was Barbara Eden. It's been many years since Eden starred in the sitcom *I Dream of Jeannie*, but the years have treated her well; and during that time she has often been on my mind. Oh, it's not what you think — my interest was generated by a different kind of chemistry.

Like any good genie, Jeannie appeared when her master rubbed her bottle the right way. A jet of smoke would gush from the bottle, and there she'd be. This splendid bit of chemical magic was accomplished with the help of hydrogen peroxide, a chemical that most people know as a disinfectant. Hydrogen peroxide has an intriguing property. It can decompose to yield oxygen and water and produce a great deal of heat in the process. Under ordinary conditions this reaction proceeds very slowly, but it can be speeded up in various ways. Light will do

it, and that's why the substance is always sold in brown bottles. And various chemicals can also catalyze the reaction. A catalyst is simply a substance that speeds up a reaction without being consumed in the process. For demonstration purposes, the classic catalyst used to decompose hydrogen peroxide is manganese dioxide, the black powder found inside batteries. If you add a bit of this to the peroxide, a cloud of white smoke almost instantly appears. The "smoke" is actually a cloud of tiny water droplets, which forms when water that is vaporized by the heat of the reaction condenses. It's very impressive. This is the reaction I described in the introduction to this book, and it has served me well in many a chemical magic show.

Reliable sources tell me that this particular reaction was also employed to make Jeannie issue from her bottle. One camera would focus on the smoke and another on Barbara Eden. As the smoke shot faded into the one of Jeannie, it would seem as if she were actually coming out of the bottle. I have often told this story during my performances, accompanied by the relevant demonstration, never dreaming that I would one day meet Jeannie herself. So now you understand my excitement when I discovered that she was there in Vancouver, at the booth next to mine. When I spoke to her she told me that she wasn't familiar with the chemistry of the genie bottle — in fact, she couldn't recall how she had managed to get out of the bottle.

I'm not positive that manganese dioxide was the catalyst that the makers of *I Dream of Jeannie* used, but it's a good bet that it was. There are, however, other substances that could have worked. Like raw liver. Catalase, an enzyme found in liver, readily decomposes hydrogen peroxide. This may seem like a curiosity, but it isn't really so strange once we consider that hydrogen peroxide is found in the human body. It is a byproduct of metabolism, and not a particularly desirable one, because hydrogen peroxide in the body is a source of hydroxyl free radicals, which have been linked with effects ranging from aging to cancer. Catalase, along with another enzyme, glutathione peroxidase, ensures that concentrations of peroxide do not run amok by converting peroxide to water and oxygen. Heat destroys these enzymes, so for the genie demonstration cooked liver just will not do.

Our blood, as well, has to deal with the effects of hydrogen peroxide. In addition to catalase, hemoglobin, the molecule that carries oxygen around our system, has peroxidase activity. In other words, it, too, can break down hydrogen peroxide to water and oxygen. This bit of chemistry aids in the common test for occult blood in the feces. Blood in the feces may be a sign of a serious problem, but it's often invisible. The physician testing for it will usually apply a small bit of fecal matter to a filter paper impregnated with a reagent that turns blue if blood is present (sometimes one wonders why so many students want to go into medicine!), and then here's what happens. The reagent contains colorless tetramethylbenzidine along with strontium peroxide, which forms hydrogen peroxide under the experimental conditions; if blood is present, the hydrogen peroxide is converted to oxygen, which in turn reacts with the tetramethylbenzidine to create a telltale blue compound. When this happens, a colonoscopy isn't far behind.

Hydrogen peroxide not only catches problems in the colon, but it can also catch criminals. When the police arrive at a crime scene and discover what they suspect are blood stains, they can test them using a procedure similar to the one I described for fecal blood. But what if the perpetrator has tried to clean up the evidence? All is not lost, as long as the police have access to a reagent that is prepared with hydrogen peroxide and a fascinating compound called luminol. When this compound reacts with oxygen, it produces a beautiful green light, a phenomenon called chemiluminescence. The crime scene is sprayed with the reagent, and the room is darkened. Wherever there are traces of blood, hemoglobin catalyzes the breakdown of hydrogen peroxide to oxygen, which then reacts with the luminol. You can't miss the glowing evidence.

The phenomenon once enabled Chicago police to apprehend a murderer. Alerted by neighbors, they questioned a man whose wife hadn't been seen for a week. He professed his innocence, but investigators squirted luminol reagent on the carpet of the couple's home, switched off the lights, and witnessed the glow. The man was so shaken by this that he broke down and confessed within an hour. I suspect he didn't appreciate the fascinating chemistry that had occurred between hydrogen peroxide, hemoglobin, and luminol.

Hydrogen peroxide can even catch microbial criminals, like those that produce the green iridescence sometimes seen on ham or other cold cuts. Bacteria, as do humans, produce peroxide, which attacks myoglobin, the reddish pigment in meat.

When myoglobin breaks down it yields some pretty unappetizing green compounds. See for yourself — just sprinkle some hydrogen peroxide on your baloney. Speaking of baloney, there is no credible scientific evidence that using hydrogen peroxide internally has any effect on cancer or aids, a claim made on numerous Internet sites.

There is, however, some truth to the story that during World War II the Germans used hydrogen peroxide in an ingenious fashion. The V-2, the rocket that terrified England, could not have functioned without the hydrogen peroxide/manganese dioxide reaction. The "genie" was used to power the pumps that delivered the fuel and the liquid oxygen to the combustion chamber. The Germans also developed a top-secret airplane, known as the Komet, that employed hydrogen peroxide to oxidize a mixture of hydrazine and methanol. This produced the hot gases that propelled the vehicle. The Komet could climb to the then-incredible height of 33,000 feet.

After the war Bell Aerosystems in the United States experimented with a rocket pack that could be strapped to a person's back. The contraption had a fuel tank containing hydrogen peroxide, which produced steam when mixed with a catalyst. A fiberglass corset protected the rocketeer, who could be propelled as far as 360 feet at thirty miles per hour as the steam rushed out through a pair of cleverly designed nozzles. A flying genie, you might say.

Appearing Aliens and Disappearing Urine

"Grow an alien!" That's what the little package said. Figuring that this could be the only way I would ever get to see an extraterrestrial, I bought it. Following the accompanying instructions, I immersed the plastic packet in a large bowl of water and waited. The next morning I was greeted by a little green man with huge bug eyes, straight out of *The X-Files*. It was quite magical, but quite terrestrial. Actually, it was polyacrylical.

Let's start our investigation of alien anatomy with a very down-to-earth story. A young mother took her twelve-month-old baby girl to the hospital to be operated on for a urinary-tract irregularity. Afterwards the doctor instructed the mother to monitor her baby's urine output to make sure the surgery had succeeded. She was to count the wet diapers. About two weeks after the operation she panicked when the baby did not void within one twenty-four-hour period. The doctor immediately ordered an ultrasound to determine whether the baby's bladder was distended, but everything appeared normal. The baby had obviously been peeing, but where was the urine?

The mystery was soon solved. The mother had switched to a new type of diaper, labeled "superabsorbent." When she checked for wetness and repeatedly found a dry diaper, she assumed that her daughter had not been urinating. What she did not realize was that a superabsorbent diaper will hold up to three hundred milliliters of urine — twice the capacity of a regular diaper — before it feels wet to the touch. When the discarded diapers were weighed, it became obvious that everything was fine. The material inside the superabsorbent diapers responsible for this nifty piece of chemical magic is popularly referred to as "super slurper"; in chemical circles it is known as a cross-linked polyacrylate polymer. Neat stuff, indeed. Some versions can retain five thousand times their own weight in distilled water, and the scope of their use extends far beyond the nursery.

Superabsorbent polymers were first described in 1938, when polymer chemistry was still a fledgling field. Chemists had discovered that small molecules could be linked together to form giant chains, or polymers, which had decidedly different properties from their unlinked components. One of these was sodium polyacrylate, a dry white powder that had amazing properties. When exposed to water, it acted like a sponge and formed a gel. Although the chemistry of this absorption is complex, the effect hinges on the high sodium concentration inside the polymer network. Water rushes in to dilute the sodium concentration (a phenomenon known as osmosis) and gets trapped inside. The effect can be reversed by adding salt (sodium chloride) to the gel; this makes the water flow in the opposite direction, converting the gel into a liquid.

Super slurper has wondrous properties, and it amazed people when it was first demonstrated. For some time no one seemed to know how to exploit its potential, then in 1968 chemists came up with the idea of incorporating polyacrylates into diapers. Here was a way to keep urine away from a baby's bottom and reduce diaper rash. The Japanese, always at the forefront of high-tech innovation, ate it up. Not literally, of course, but they could have — toxicity tests showed that the material was remarkably safe. By 1984 North Americans had also discovered the advantages of superabsorbent diapers, and the product rapidly dominated the market. Soon feminine hygiene products were also incorporating the new technology, as were adult incontinence products.

Advances came quickly. Chemists designed special acrylic polymers that could be used in the manufacture of leak-proof tape to wrap around undersea cables. Or blended with rubber to produce a mortar that swelled on contact with water, providing a tight seal. This was used in the construction of the Chunnel, the artery that connects England and France beneath the waters of the English Channel. You sure wouldn't want any water seeping through those walls. Super slurpers have also proven themselves useful in the little pads you sometimes find under meat packaged for sale at your local supermarket; they absorb the fluids released by the meat, making for a safer and more appealing product. Polyacrylates are even used in the filters that remove water from gasoline.

Farmers, too, have profited from this technology. Corn seeds coated with slurper in the form of polyacrylamide can now be planted under dry conditions, because when watered they will retain the moisture. In one experiment, coated seeds produced far more plants per acre than uncoated seeds. The stuff can even be mixed into the soil to help moisture retention and prevent erosion. Watering frequency on lawns, golf courses, and irrigated crop fields can be reduced by as much as forty percent. Wouldn't it be great if you didn't have to ask the neighbors to water your plants while you're away? You don't; just poke a few holes in the soil of potted plants with a pencil and fill with slurper. Add water, and you're good for a couple of weeks. But don't overdo it. If you add too much polyacrylamide to the soil, it can swell enough after watering to break the pot, and what a slippery mess that makes. Indeed, this is the only real downside of super slurper. If you step on the gel, down you'll go.

Just when you think that we've explored all the possible uses for polyacrylates, up pops a new idea. How about using it to battle fires? In the early 1990s Florida firefighter John Bartlett was rummaging through the debris of a fire when he made an interesting discovery. His attention was drawn to an item that had not been consumed by the flames — a used diaper. What had prevented it from burning? The inquisitive Bartlett investigated and soon discovered that the polyacrylate inside had swelled with water and acted as insulation against the flames. This gave him an idea, and he got down to work. After five years of effort, Bartlett finally succeeded in transforming the powder into a gel that has the appearance of shaving cream. During a fire the substance can be applied with a hose over virtually anything, preventing it from burning. The Los Angeles Fire Department already uses it, as do the Florida Power and Light Company and the U.S. military. A great invention, but apparently not an original one. Someone registered a patent in 1966 to use polyacrylate as a water-immobilizing gel in firefighting. John Bartlett may have done no more than reinvent the wheel, but he is undoubtedly the one who got it rolling in the right direction.

Now back to our alien autopsy. As it turns out, the creature is just a sophisticated super slurper. His flesh is made of acrylic polymers (partially hydrolyzed acrylonitrile, for those of you interested in the nitty gritty details), which have been chemically linked to starch to achieve just the right gooey consistency when swollen with water. By immersing the super slurper in a salt solution, we can produce a deflating effect. The alien withers away, but I hope his chemistry is no longer alien.

A Bang of a Discovery

"The inside just exploded with a yellow flame, and then there was soot everywhere," a shaken witness told police as he stood staring at the remnants of a car. Four men with severe burns had been pulled from the wreck, but the police were initially at a loss to explain what had happened. Slowly they pieced the story together. The car's occupants had been transporting eight balloons filled with a highly combustible mix of acetylene and oxygen. They were planning to use them to put a bang into an otherwise dull day. When they lit a joint, the day brightened up all right, but not in the way they had hoped.

I'm sure these four guys didn't comprehend the fascinating chemistry behind their adventure or the curious historical link between the balloons and what they contained. The story of the discovery of acetylene and its connection to rubber is one that merits telling, because it is such a great example of how knowledge, intuitiveness, and serendipity come together in scientific discovery.

On May 4, 1982 Thomas Willson, a Canadian inventor, placed some calcium oxide (lime), coal tar, and aluminum oxide in a container and heated the mixture to a high temperature in a furnace. He was hoping to produce metallic aluminum, which at the time was an extremely expensive commodity. His thinking was good. Coal tar, which was basically carbon, was known to convert lime to metallic calcium. Willson knew that calcium was a more active metal than aluminum, and he hoped that it would strip oxygen away from the aluminum oxide. After heating his solution for some time, he opened the furnace, hoping to see shiny metallic aluminum, but what he saw was a dark residue. Frustrated, he threw the mess into the stream that ran past his lab. This, of course, was long before environmental laws had come into existence, and that was a lucky thing.

As soon as the residue hit the water, huge bubbles began to form, and a plume of water shot up into the air. In an instant Willson's disappointment changed to elation. Whatever it was he had made, it seemed more interesting than the aluminum he had sought. Willson repeated his experiment, and to his great relief found that he was able to reproduce the residue that produced the gas when it reacted with water. Furthermore, the gas burned with a bright, sooty flame. Chemical analysis showed that Willson's residue was calcium carbide and the gas was acetylene. But that highly combustible acetylene was not a new discovery; thirty years earlier Friedrich Wohler, the famous professor of chemistry at the University of Gottingen, had made calcium carbide by heating calcium with charcoal to a high temperature, and he had observed that it formed acetylene when it reacted with water. This was not an efficient way of making the carbide, however, so it wasn't a good method of creating the gas. Willson's method yielded large amounts of calcium carbide easily and was therefore practical for making acetylene.

Why was this important? The 1890s was the era of the gaslight. The world's first gaslight company was established in London in 1813, and subsequently the city had installed an elaborate network of gas lines to ensure that streets and homes were well lit. But mobile lighting was still restricted to candles and kerosene lamps. Willson realized that his acetylene, which burned with a far more brilliant flame than kerosene, had great market potential. By 1895 he had founded the company that eventually became Union Carbide, one of the biggest chemical companies in the world. Soon consumers were able to purchase lamps based on calcium carbide, clever devices in which water dripped into a container of carbide and generated acetylene gas. This gas, in turn, flowed to a nozzle where it could be ignited. A mirrored surface behind the flame increased the intensity of the light.

Car manufacturers jumped on the idea and used carbide lamps for headlights. Miners also used the lamps, although not without risk. Combustible gases are often present in mines, and in times past these gases were sometimes ignited by carbide lamps, with tragic results. Today carbide lamps have been relegated to the museums, but acetylene is still one of the most important industrial chemicals in existence. In fact, modern life would not be possible without acetylene. In 1895, the same year that Willson established his company, the French chemist Henry-Louis Le Chatelier, professor of chemistry at the Collège de France, discovered that when we burn acetylene with an equal volume of oxygen, the temperature of the flame, over 3,000°C, is far higher than that achievable with any other gas. The flame Le Chatelier produced was hot enough to melt steel, and the concept of welding was born. Without oxyacetylene welding torches, construction as we know it would not be possible.

Acetylene has yet another use. About half of all acetylene produced today goes towards the production of other organic chemicals. Adding hydrogen cyanide to acetylene, for example, yields acrylonitrile, which is used in the production of acrylic fibers. Acetylene can also be converted into vinyl acetylene, which is the raw material needed for the manufacture of neoprene, one of the most useful synthetic rubbers.

This development goes back to the 1920s. Chemists had already discovered that the stuff oozing out of the rubber tree could be broken down to a simple molecule called isoprene. Since they could manufacture isoprene in the lab, might it not be possible for them to create the reverse reaction and make synthetic rubber? A shrewd thought, but no matter how hard chemists tried to string together these isoprene molecules they could not produce an acceptable rubber. Then Wallace Carothers, the DuPont chemist who would later discover nylon, had an idea. He knew that Reverend Julius Nieuwland, a chemistry professor at Nôtre Dame, had developed a deep interest in Willson's acetylene and had reacted the substance with hydrochloric acid to make something he called vinylacetylene. The chemical was highly reactive and had a formula very similar to that of isoprene. Carothers thought that it could be converted into some sort of rubber.

He gave the project to an assistant, who tried all sorts of reaction conditions — to no avail. Then one weekend he left one of his experiments sitting in a flask and discovered on Monday morning that it had turned into a rubbery solid. Analysis showed that residual hydrochloric acid had reacted with the vinylacetylene to make a synthetic rubber. Neoprene was born, and what a substance it was. It was resistant to oil and gasoline and was the ideal material for engine belts and gasoline hoses — far better than natural rubber. Neoprene was also great for making huge balloons, like the ones used in Macy's Thanksgiving parade. Furthermore, it gave chemists the impetus to develop other synthetic rubbers, like the ones that may have been used in the making of our acetylene adventurers' balloons.

These four unfortunate guys may be interested in yet another acetylene connection — one that involves the idea of getting stoned. French researchers have found a way to reduce crop damage caused by hail. They have developed a gun equipped with a combustion chamber in which acetylene mixes with air to create an explosion; the gun's hornlike barrel is aimed upward to send a shock wave into the clouds. A series of such shock waves breaks up the hail stones, ensuring that the crops below — and people — don't get stoned.

By All Indications

Sometimes I like to take a glass of water and turn it into wine. This standard chemical trick relies on a little knowledge of acids, bases, and indicators. If you have ever added vinegar to purple cabbage juice and watched as its color changes to red, then you've experienced the wonders of an indicator. And if you have never carried out this epic experiment, then I would urge you to do so now. Next, add some baking soda and amaze yourself by turning the red cabbage juice green. Indicators, simply put, are substances that are one color in an acid solution and a different color in a base. The best-known substance of this kind is probably litmus. Remember dipping little pieces of colored paper into an unknown liquid to determine whether the liquid is an acid or a base? That litmus paper you used was impregnated with an extract of lichens that turns red in acid and blue in base. But the water-into-wine trick requires a different indicator: it calls for phenolphthalein.

This chemical with the tongue-twisting name was first synthesized in 1880 as a derivative of phenol, an antiseptic that had just come into common use. Phenolphthalein itself never made it as an antiseptic, but it did prove to have an unusual property — it was colorless in an acid solution and a beautiful bright pink in base. So now you know the secret behind the wine trick. Just add a little base to a glass of water (I use potassium hydroxide, but baking or washing soda will do) and pour it into a glass containing a touch of phenolphthalein. The ensuing color change will seem miraculous. I've even seen an attempt to capitalize financially on this bit of chemical flim-flammery. The novelty item "bloody soap" is supposed to strike terror into the hearts of unwitting victims who wash their hands with it.

The bar of soap looks ordinary enough, but it is treated with a little base and some phenolphthalein, which combine when the bar is moistened, leaving the hands "dripping with blood." I assume the product's creators intended to generate psychological torment for the victim and fun for the perpetrator. I really wouldn't bother, though. "Dripping with blood," is not a realistic description of the effect; "trickling with dilute strawberry Kool-Aid" is more like it. But, hey, who says science has to stand in the way of entertainment? Surely not MacGyver. Remember him? MacGyver was the TV hero who fought criminals with his brains instead of brawn. I liked him. He wasn't some frizzy-haired chap with thick glasses dressed in a stereotypical white lab coat and sporting a plastic pocket protector filled with more pens than anyone could use in a lifetime. He was a regular guy who knew how to use chemicals — like phenolphthalein — to his advantage.

In one classic episode, our hero tries to escape on foot from the hail of bullets some nasty guys are unleashing in his direction. Unfortunately, he's hit in the shoulder and leaves a trail of blood for the nasties to follow. Seeking refuge in a building that turns out to be an abandoned hospital, MacGyver discovers some interesting chemical solutions to his problem. He rifles through a few cabinets and discovers a box of x-Lax. Joy reigns supreme! This is a curious reaction, indeed. You would think that a laxative would be the last thing on the mind of a man running for his life. But to MacGyver, x-Lax isn't merely a laxative — it's a way out of a jam. He knows that its active ingredient is phenolphthalein. That's right, our old acid-base indicator friend also wears another hat.

MacGyver quickly hatches a scheme, but he's missing an ingredient. After some more frenetic running around and cupboard-opening he finds a can of drain cleaner under a sink. And MacGyver knows that this is actually sodium hydroxide, a potent base. He swiftly combines the x-Lax and the Drano to generate some fake blood. Our clever hero uses it to lay down a false trail and lure his pursuers into a trap. These criminals are so dim-witted that they mistake the deep pink of phenolphthalein for blood. But at least the show's viewers learn some neat chemistry. Don't count on duplicating this scenario, though, if you find yourself in a similar predicament. Phenolphthalein was removed from the market as a laxative because of allegations that it caused cancer in test animals that consumed it in high doses.

While we can no longer rely on phenolphthalein to get our systems moving, we can still use it to get criminal investigations moving. Remember the stir caused by the stains on the Ford Bronco and the glove in the O.J. Simpson investigation? Were they blood, or were they, as the defense suggested, just taco sauce? Criminal investigators presented a strong case for blood, and they based that case on phenolphthalein. Here's the story. When we heat phenolphthalein with zinc, it converts into a compound (reduced phenolphthalein) that does not produce a pink color with base. But we can turn this derivative back into phenolphthalein by infusing it with sufficient amounts of molecular oxygen. Such oxygen can be generated from hydrogen peroxide in the presence of a catalyst. Blood has just such a catalyst. Hemoglobin, found in all red blood cells, is very adept at hastening the decomposition of hydrogen peroxide into oxygen. This is a very important function, because hydrogen peroxide, a by-product of metabolism, is quite capable of damaging tissues.

So investigators can treat a suspicious stain with some reduced phenolphthalein and hydrogen peroxide. If it is indeed blood, the hydrogen peroxide will release oxygen, which oxidizes the reduced phenolphthalein to phenolphthalein. Investigators then treat this with base, and if the characteristic pink color appears — bingo! And did this happen with the O.J. stains? It sure did, but the defense argued (quite effectively, as it turns out) that certain fruits and vegetables also have enzymes that can liberate oxygen from hydrogen peroxide. Simpson's lawyers contended that taco sauce would have yielded the same result as blood. Actually, plant enzymes break down easily while the required blood enzymes do not. The investigators in the Simpson case could have turned up the heat on O.J. by applying heat to the sample, but they just tested for cold blood.

Phenolphthalein did not solve the Simpson case, but it could have prevented an expensive escapade that occurred on a New Jersey highway. During a traffic accident an air bag inflated and the car filled with a fine white powder. Arriving on the scene, the rescuers, fearing that a toxic substance had been released, cut off the driver's clothes and cordoned off the area. They even set up a shower for anyone who may have come into contact with the powder. Eventually, emergency workers took twenty-two people to hospital for observation. Why were they so concerned? Because caustic sodium hydroxide (lye) can form as a by-product of the chemical reaction that inflates an air bag and officials at the accident scene thought that the white powder released was sodium hydroxide. It actually turned out to be talcum powder, a lubricant for the air bag. Authorities could easily have confirmed that the powder was not sodium hydroxide by using phenolphthalein.

Phenolphthalein is fascinating, isn't it? Now you understand why I enjoy turning water into "wine." And no, for my next trick, I will not attempt to walk on water.

The Magic of Secret Inks

I was going through some of my old papers and I came upon an item that I'd long since forgotten about. It was an ad for my very own Prof. Schwarcz's Miracle Color Changing Water Wand. I billed my magic wand as "Revolutionary! Mysterious! Automatic!" with the power to turn water black and then, at the magician's command, restore it to a colorless state.

In the 1970s, when I designed this little magic curio, I thought I was pretty clever. I based my wand's "magic" on the fact that gallic acid reacts with iron chloride to form a dark blue-black complex that can be destroyed by citric acid. (The magician's code of honor prevents me from disclosing the exact mechanics, but I'm sure you can make a reasonable guess about the workings of the wand.) It now seems, however, that I was about two thousand years too late in putting these chemical reactions to a novel use.

Philo of Byzantium, a Greek military scientist, developed the world's first secret-ink recipe using the same type of formula.

He noted that an extract of nut galls (the little knobs that grow on the trunks of oak and nut trees) turn a dark blue-black in the presence of certain iron compounds. Philo probably observed this effect when he added minerals or clay with a high iron content to the extract, and that sparked a brilliant idea. Why not write on paper with the colorless gall extract and "develop" the writing with a mineral solution? It worked, and secret inks came into being. This interesting color chemistry can actually be observed in an upper New York State waterway called Black Brook. Iron deposits from the banks of the brook combine with gallic acid draining from peat swamps to produce "ink."

In general, secret inks derive from two reagents that produce a color. Sometimes one of the reagents is heat — the simplest ink is made by writing on paper with a solution of a colorless substance that chars when it is heated. Lemon juice works well for this purpose, and it was apparently used during the French Revolution. Urine does the trick, as well. Many a message has been sent from one penitentiary inmate to another via this unusual method.

One of the simplest chemical combinations for use in secret messages is phenolphthalein and ammonia. A message written with a solution of the indicator phenolphthalein is colorless, but it turns pink when exposed to ammonia vapors. A German spy, George Dasch, exploited this technology during World War II. Dasch sneaked into the United States after a submarine dropped him off at a point along the Atlantic coast, but someone saw him and he was apprehended. Authorities found no incriminating evidence on Dasch, but they became suspicious about the number of white handkerchiefs he was carrying with him. At the FBI laboratory chemists solved that mystery. Dasch had listed his contacts and addresses on the hankies with phenolphthalein. In court his prosecutors demonstrated this using ammonia, and Dasch was executed. Espionage in wartime is a capital crime.

Perhaps the most spectacular secret ink is Prussian blue, which forms by means of a chemical reaction between ferric sulfate and potassium ferrocyanide. Generally, a message written with ferric sulfate solution will be revealed when it is sprayed with ferrocyanide. A spy can soak fabric with each of these solutions and transport secret information without detection. During World War II a German spy named George Vaux Bacon made notations on his socks and cloth buttons with the secret ink reagents. He, too, was caught and executed.

Another secret-ink system used by the Germans during the war involved the chemical reactions between lead nitrate and sodium sulfide. As any high school student should know, solutions of these compounds are colorless, but form a black precipitate of lead sulfide when mixed. A spy can inscribe a message with the lead solution, and its recipient can read the missive by spraying it with the sodium sulfide solution.

Oswald Job, another German spy, was sentenced to death in 1944 when the invisible chemicals were discovered in a hollow key he was carrying in his pocket.

Even in this age of high technology secret inks may not yet be passé. Recently, a most unusual case surfaced. According to reports, Egyptian authorities arrested two men on suspicion of spying. The evidence? They were in possession of a pair of ladies' underwear with markings in secret ink that they had been passing back and forth between them. If the case ever comes to trial, then its outcome may well hinge on the chemistry of color-changing underwear. Sounds kind of interesting. Maybe there's a market out there for Prof. Schwarcz's Miracle Color Changing Water Wand that needs to be explored. Might be just the thing for secret agents.

Blast from the Repast

I had never really thought much about the behavior of dead shrimp — at least, not until I heard that they were harassing people. I was alerted to this strange phenomenon when someone asked me a question about some oven-cooked shrimp that had jumped off the plate into the face of a diner. I must admit that I had a hard time picturing the scenario. The shrimp were arranged nicely on a plate when they unleashed their attack. They took off, flying wildly, leaving a trail of melted butter in their wake. My first suspicion was that the shrimp dinner in question had been accompanied by a little too much wine, but I was wrong. Several other gourmets later reported to me that they, too, had seen cooked shrimp exploding off their plates like popcorn.

Now I had enough evidence to begin an investigation, and the "popcorn" description provided me with a vital clue. What if the natural water content of the shrimp had turned to steam in the oven and had been unable to escape through the shell? The little sea creatures would then become potential steam bombs. A slight disturbance could conceivably crack the shell and release the steam. Newton's third law would quickly come into play: for every action there is an equal and opposite reaction. As the steam shot out of the shrimp, they would become veritable projectiles. It was kind of like bursting an overinflated balloon. Or popping corn. That's right — corn pops because there is a tiny bit of moisture inside each kernel that changes into steam when heated. The expanding steam has no place to go and literally causes the kernel to explode. Now I had my theory; all that I needed was the confirming experiment. So I headed off to the lab — that is, to the kitchen.

Setting the oven to 350°F, I buttered and garlicked the shrimp and placed them in their little custom-made dish. I didn't have to wait long for the results. The first shrimp launch took place inside the oven, splattering shrimp shrapnel everywhere. The other shrimp had the decency to wait until being served before blasting off. Now, I don't want to mislead you and create the impression that my kitchen was filled with flying shrimp. Most of the little carcasses lay quietly on their serving plate. But there was the odd bloated one that mustered its energy and, with a hiss of escaping steam, avenged itself on the chef. The mystery of the jumping shrimp was obviously caused by a gas explosion.

With that problem solved, we can now take up the case of the exploding Swiss cheese. Characteristic of this type of cheese are the large holes produced by the carbon dioxide gas that forms during the aging process. Cheese makers add a variety of bacteria to their creations as a way of ripening them. To Swiss cheese they also add a strain of *Propionibacter shermanii.* This bacterium uses the lactic acid secreted by other microbes as food and gives off huge quantities of carbon dioxide gas, which forms holes as it expands. Another by-product of this reaction is propionic acid, which contributes to the flavor of the cheese. For a very long time Swiss cheese makers happily supplied the world with their holey products — but then something went wrong.

Huge wheels of Swiss cheese, worth hundreds of dollars each, began to explode in storage. An investigation uncovered the fact that the Swiss Milk Research Institute, in its never-ending quest to improve cheese production, had added a new strain of aggressive bacteria to the cheese culture and neglected to inform the cheese makers of this breakthrough. The only thing that broke through, however, was the carbon dioxide. So much pressure built up inside the cheeses that they blew themselves to bits. But you need not glance warily at that wedge of Swiss cheese sitting in your fridge. It won't attack you. Its bacterial cultures have been readjusted to produce just the right amount of carbon dioxide.

A similar problem recently plagued chocolate-egg sellers. The Easter confections began to explode on store shelves. They had become contaminated by a mold that produces carbon dioxide from sugar; since the eggs were filled with jelly, the mold had plenty of sugar to feed on. But the mold was of a harmless variety, and the most it could do was produce a chocolate-coated consumer.

Snowboarders in Utah's Little Cottonwood Canyon may be especially interested in the exploding Swiss cheese and Easter egg episodes. Some of them have been carrying out their own little experiments with the explosive potential of carbon dioxide. Avalanche is a constant danger in the area. In order to trigger avalanches at a time when no one is around, local experts detonate explosives. Some snowboarders, who love to explore uncharted territory but do not relish being swept away by an avalanche, have reportedly developed their own avalanche-starting technique. They place some dry ice (solid carbon dioxide) in a one-liter plastic bottle and carry it with them to the top of the run. If they encounter a suspicious slope they urinate into the bottle, quickly screw on the cap, and toss the bottle into the snow. The warm urine causes the solid carbon dioxide to vaporize, pressure builds up, and soon there is an explosion that they hope will eliminate the avalanche threat. Not likely, says the National Avalanche Center. The bang you get for your urine is not enough to trigger an avalanche, and this imaginative strategy only leads to a false sense of security.

So, gas buildup accounts for a range of interesting effects, but none is more bizarre than exploding human bodies. Let me warn you: this account is not for the squeamish. Not long ago an insurance company had to deal with a most unusual problem — an indescribable odor in the apartment of a client. The source? A dead body that had "exploded" eight days earlier, scattering fragments everywhere. The apparent cause was gas gangrene caused by bacteria of the *clostridia* variety, which are found in the soil as well as in the intestinal tracts of humans and animals. These don't usually cause a problem — at least, not until they become active under the right conditions. *Clostridia* bacteria multiply only in the absence of oxygen, but when they do they produce a variety of toxins and gases. A typical example would be a battlefield wound. Blood does not circulate through crushed tissue; that tissue is therefore deprived of oxygen. If *clostridia* bacteria get into the wound, they become active and produce gas gangrene. About ten percent of all World War I battlefield wounds developed gas gangrene, but the incidence was far lower during World War II, due to the availability of better treatment techniques.

Obviously, blood does not circulate through a dead body. In rare instances so much gas builds up inside a corpse that the skin bursts open, resulting in a very nasty scene. This is exactly what confronted the insurance company. How did it solve the problem? By calling in an outfit that specializes in removing smells with an ozone generator. The machine converts oxygen to ozone, which then oxidizes smelly compounds.

To tell you the truth, when I set out to solve the mystery of the exploding shrimp, I never thought I would end up discussing gas gangrene. But that is one of the fascinating aspects of science: you initiate an investigation and you never know where you will end up. Oh — the solution to the shrimp problem? Just pierce each one with a needle before baking. The steam will vent through the hole and the explosion risk will be eliminated.

Out of the Blue, in Prussia

The stench in the alchemist's laboratory must have been awful. Bubbling kettles filled with blood, bones, and other animal remains spewed their foul vapors. In another corner vegetable matter and wood ashes rotted away slowly, yielding potash. An unpleasant working environment to be sure, but it was a necessary condition of a trade devoted to unlocking the secret of life, unearthing the key to longevity. Johann Konrad Dipple was seeking nothing less in the early years of the eighteenth century. Holed up in Berlin's Wittgenstein Castle, he finally manufactured Dipple's Oil by distilling animal offal over potash. What an acrid taste and a repulsive odor that concoction must have had! Strangely, that was enough to guarantee its success: people believed that anything that tasted and smelled so bad had to be good for you. Dipple's fame spread, and his universal cure sold for a hundred years.

One day a color merchant named Diesbach came to see

Dipple in his Prussian castle. Diesbach was not seeking longevity. He'd come to ask for help with his dyestuffs — an early chemical consultation, one might say. Herr Diesbach had some experience with making a red dye from the cochineal insect; he cooked the little bugs up with "green vitriol" and potash. Now, however, he was having trouble finding potash. The stuff he referred to as green vitriol was iron sulfate, and potash was the "ash" left behind when a mixture of wood residue and vegetable matter was boiled to dryness in a "pot." It was mostly potassium carbonate. Dipple had plenty of potash — after all, it was a key ingredient in his magical remedy. But when Dipple's potash was mixed with Diesbach's vitriol, the results were absolutely startling. A beautiful blue color appeared, and they christened it "Prussian blue." The pigment quickly replaced the naturally occurring aquamarine in inks and paints, because while aquamarine did provide a beautiful blue color, it was very expensive. (Not surprising, since it was derived from emeralds.)

Neither Dipple nor Diesbach understood the chemistry of their accidental discovery, but it was clear that it had something to do with the potash being contaminated with animal residue. Factories sprang up all over Prussia, and they used Dipple and Diesbach's formula to crank out tons of the new dye. Reeking of hides and blood, these factories were hellish places. But the smell wasn't the worst of it. The screeching of iron stirring paddles against iron kettles sent visitors to those factories scurrying towards the exits.

Paying a visit to a Prussian blue factory, Justus von Liebig, professor of chemistry at the University of Giessen, asked the owner why he'd done nothing to diminish the deafening noise. "Why, Herr Professor," he replied, "we get the most blue pigment when the screeching is the loudest." Liebig pondered this observation and finally suggested that adding a handful of iron filings to the dye mixture would reduce the amount of stirring and screeching required. Sure enough, when the factory workers followed his instructions they were able to produce the brilliant blue color without the vigorous stirring.

No one understood why the chemist's brilliant notion worked until Prussian blue was identified as ferric ferrocyanide, a compound essentially composed of iron and cyanide. The discovery was fortuitous. Proteinaceous animal waste contains plenty of nitrogen and carbon, which reacts under heat to produce cyanide. In the presence of ferrous ions (a form of iron) and potash, cyanide then forms a complex substance called potassium ferrocyanide, an unimpressive, pale yellow compound. But it reacts with ferric ions to form ferric ferrocyanide, or Prussian blue, so a lot of things had to go right for Dipple and Diesbach to make their discovery. The potash had to contain cyanide and the green vitriol had to be contaminated with some ferric ions. Vitriol, or ferrous sulfate, is commonly contaminated in this manner, since it reacts with oxygen in the air to produce ferric sulfate. Dipple and Diesbach certainly lucked out when they produced Prussian blue, and they were clever enough to capitalize on their discovery. We can now explain Liebig's contribution, as well. The fine scrapings produced by scratching the iron provided plenty of surface area to react with oxygen and produce rust or, chemically speaking, ferric ions.

Prussian blue, in the form of printers' inks, artists' colors, and paints, soon flooded the market. It also stimulated interest in other potentially useful substances that might be derived from the potash, iron, and animal-residue mixtures. One of these, known as "red prussiate of potash," did turn out to be very useful. When combined with ferric ions, it didn't produce a dramatic color until it was exposed to direct sunlight. Then it turned blue. Prussian blue. The discovery revolutionized architecture. Blueprints were in the offing. No longer did architects have to reproduce their drawings by hand. They could now draw on transparent paper and then place their drawings over sheets sensitized with the appropriate light-sensitive chemicals. When exposed to light, the sensitized paper turned blue everywhere except where pencil lines prevented the light from shining through the drawing. The blueprint was essentially a negative, with the architect's lines in white. As many copies as needed could be made. Blueprints held sway until they were replaced by another chemical marvel: xerography.

The fact that cyanide was a critical ingredient in Prussian blue generated interest and concern. Was handling Prussian blue dangerous? As it turns out, no. Cyanide, of course, is highly toxic, but in Prussian blue it is tightly bound to iron and is chemically inert. While Prussian blue doesn't cause cyanide poisoning, it can detect it. If we suspect that cyanide is present in a tissue sample, we add the ingredients needed to make Prussian blue (now known to be ferrous hydroxide and ferric chloride) to force the cyanide to declare its presence. When a number of horses died suddenly on a Tennessee farm, the farmer's explanation that they had been struck by lightning seemed implausible. An investigator noted that the animals' blood was cherry red, a sure sign that it contained plenty of oxygen that was not being passed to the cells. What could cause this? Cyanide, for one thing. It destroys an enzyme that cells need to use oxygen. Sure enough, when the investigator applied cyanide detection reagents to tissue samples, the unmistakable Prussian blue color appeared. Where had the cyanide come from? Hay. Many plants contain naturally occurring cyanides but rarely present a problem because the cyanide is not in an absorbable form. However, damage to plant tissues can release an enzyme that liberates the cyanide. This is precisely what happened on that Tennessee farm.

If Prussian blue can be used to detect cyanide, which is one of its ingredients, why not iron, another one? There is no reason why it can't. When the right reagent containing cyanide is added to a sample containing ferric ions, Prussian blue forms readily. This is the basis of a laboratory test for detecting iron deposits in the human liver. The formation of Prussian blue upon addition of a reagent to the biopsy sample indicates iron overload, a potentially serious condition.

Many people owe a debt of gratitude to Johann Dipple. I'm one of them. For many years I have performed a chemical magic trick that consists of spraying a blank sheet of paper with a yellowish solution to reveal a blue message or drawing. I pretreat the paper with potassium ferrocyanide and then spray it with ferric sulphate to produce the stunning color of Prussian blue. And what eventually happened to Dipple? In 1734 he was found dead in his laboratory, clutching a new version of his life-extending formula, which he'd just tasted. He should have stuck to producing Prussian blue.

Roots and Wings of Red

"The British are coming, the British are coming!" cried Paul Revere. Well, he could hardly have missed them, decked out in their bright red coats. The attire of the British soldier tended to undermine military strategy, but the color had nothing to do with battle tactics. It had to do with economics. The British uniforms were required, by royal decree, to be dyed red in order to support British agricultural interests, specifically the cultivation of the madder plant (*Rubia tinctorum*). It was from the roots of this plant that the brilliant red dye known as alizarin was extracted. Alizarin was not produced in the colonies, and

American soldiers had to settle for blue uniforms, dyed with indigo. "Blue-bellied Yankees," the British smirked, but we know who had the last laugh on that one.

While the expression "red coat" certainly conjures up an image of the British Revolutionary War soldier, madder root as a dye was not a British invention. Thousands of years before Paul Revere's famous ride, madder was already in use. The Egyptians colored the wrappings of mummies with it. The stunning red togas of the Roman centurions also owed their color to madder juice. But perhaps the most ingenious use of madder extract can be attributed to Alexander the Great. Before a critical battle with the Persians, he ordered his soldiers to splash themselves with madder juice and stagger onto the battlefield. The Persians, thinking they could easily defeat these wounded men, attacked, thereby falling into Alexander's ingenious trap.

The madder industry was huge in Europe up to 1870, with about seventy thousand pounds of the pigment being produced annually. Then William Henry Perkin and Heinrich Caro put the farmers out of business. Perkin had already made a name for himself by synthesizing a dye known as "mauve," which had been hailed as the world's first synthetic dye. Indeed, it was exactly that, but Perkin had hit upon it by accident. He was not the least bit interested in dyes and had in fact been trying to make quinine, a much-needed antimalarial drug. One day, frustrated with his lack of progress, he emptied his chemicals into a sink and was rewarded with the appearance of a brilliant new color. Recognizing the importance of this discovery, Perkin capitalized on it commercially, and out of his success arose a lifelong interest in dyes.

During the 1860s Perkin turned his attention to alizarin, wondering if this substance could also be produced synthetically. By 1869 he was making the red dye in his laboratory. The importance of this endeavor, however, far surpassed the importance of alizarin as a colorant. This was the first time anyone had synthesized a dye in a logical, stepwise fashion based upon the knowledge of the molecular structure of the desired product. Just the year before Adolf Baeyer had shown that one could break down complicated organic molecules into simpler compounds by heating them with zinc. Alizarin, when treated in this fashion, yielded anthracene, a well-known component of coal tar.

In a brilliant piece of retro-engineering, Perkin converted anthracene to alizarin utilizing a process that he patented on June 26, 1869. Unfortunately for him, Germany's Heinrich Caro had filed a patent for a very similar process just one day earlier. So, while Perkin had the rights to alizarin in England, he missed accessing the lucrative German market by a mere twenty-four hours. Perkin's and Caro's alizarin syntheses turned out to be so practical that within a few years the price of the dye dropped from fifteen dollars to fifty-five cents a pound. Today alizarin is still used for dyeing wool, but it has been joined by a host of other synthetic dyes.

Madder may have been the most important red dye of antiquity, but it was not the most exotic one. Kermesic acid, a beautiful scarlet color, was derived not from a plant, but from a tiny scale insect. The ancient Mesopotamians had somehow chanced upon the discovery that the female kermes insect, which spends its whole life clinging to an oak tree, was capable of yielding a splendid red color suitable for dyeing garments. Collecting the insects, though, posed a problem. They had to be dislodged from the tree individually by people with specially grown fingernails. Scarlet fabrics were therefore horrendously expensive, but they so impressed Pope Paul II in 1467 that he chose kermes for the scarlet of cardinals' robes. As a result the port of Venice became a center of the scarlet dye industry, and just as the British had passed laws to protect the madder industry, the Italians attempted to safeguard the kermes market through legislation. The laws they passed were anything but wimpy. Punishment for substituting the cheaper madder root extract for kermes was severe: the offender's hand was cut off. Soon, however, the market for kermes itself would be cut off.

When the Spanish conquistadors arrived in America they were impressed with the brilliant red fabrics worn by the Aztecs. Then the Spaniards learned that the Aztecs knew of another dye-producing insect — the cochineal. A type of cactus known as *opuntia*, plentiful in South America, was the ideal breeding ground for these insects, and the females produced the dye. The Aztecs had developed the technique of sweeping the insects off the leaves into hot water to kill them before drying the corpses in the sun. There was no gender separation; the useless males were sacrificed to the fascinating chemistry of the female.

Hernán Cortés, who had arrived in America in 1518, did not fail to recognize the commercial value of cochineal. Indeed, the dried insects became the first product exported from the New World to the Old, and Europeans couldn't get enough of that vibrant, beautiful scarlet color. France's famed Gobelin tapestries were tinted with cochineal. There was such demand for the dye that by the sixteenth century Spain was importing half a million pounds of cochineal annually. That's an impressive amount, especially considering that it takes seventy thousand hand-gathered insects to make a pound of the dye. Cochineal red became so popular that it drove kermes into obscurity, and cochineal is still used today.

Dactylopius coccus insects, now bred in Peru and the Canary Islands, produce carmine, as the actual colorant is called, for the food and cosmetics industries. It is used in a wide range of items, from lipstick to maraschino cherries. Cochineal is also added to an array of foods, but not kosher products. Jewish religious law prohibits the inclusion of insects or their parts in food. Granted, others may also object to ingesting insect extract with their yogurt or ice cream, but, aside from the fact that it causes allergic reactions in a few rare cases, cochineal is a safe food colorant. So don't let the presence of a little bug juice in your Campari or your cherry ice cream bug you.

Out of the Frying Pan

The weather outside was frightful, but the product inside was delightful. So, braving a raging snowstorm, shoppers flocked to Macy's in New York City to scoop up the latest innovation in cooking technology. No more frying pans with burned-on food, was the promise; no more horribly scorched pots. The age of Teflon cookware had arrived. The two hundred people who were lucky enough to get their hands on this new marvel of science dreamed only of perfect fried eggs and pancakes. Little did they realize that the same material that now coated their pots had helped end the War in the Pacific.

In the early 1940s a top-secret project was under way at a laboratory in Oak Ridge, Tennessee. Chemists worked feverishly to separate a form of uranium, known as uranium-235, from the major naturally occurring type of uranium, uranium-238. This was not some sort of esoteric research project: it was critical to the war effort. Scientists knew that uranium-235 underwent fission when bombarded by neutrons, releasing huge amounts of energy. In simple terms, this was the key to making the atomic bomb. But uranium-235 occurred in natural uranium only to the extent of one percent, and separation involved some very sophisticated technology. Indeed, the only way to accomplish this separation was to first convert the uranium into uranium hexafluoride, a reaction that required the use of fluorine gas, a horribly corrosive substance. It was so corrosive that no material known at the time could adequately stand up to it.

In a hush-hush meeting the military asked the DuPont chemical company to present them with a material that was up to the task. Certainly, the military did not expect the DuPont people to accomplish this overnight. But somehow they did. That's because DuPont already had such a material — and it was a material that they hadn't known what to do with. In 1938 Teflon had been discovered by a young DuPont chemist named Roy Plunkett in the most curious fashion.

In 1936, having just graduated from Ohio State University, Plunkett was hired by DuPont to work on new refrigerant materials. The refrigerants used at the time, ammonia and sulfur dioxide, were highly toxic. Thomas Midgley, working for the refrigerator division of General Motors, had already searched the literature for a colorless, odorless, tasteless, nontoxic, nonflammable refrigerant and had concluded that chlorofluorocarbons (CFCs), or Freons, would be ideal. Plunkett was given the task of preparing a specific Freon, and he decided that the ideal starting material would be a compound known as tetrafluoroethylene. He prepared some of this gas and stored it in a cylinder.

The next morning Plunkett opened the cylinder valve, but, to his surprise, nothing came out. He thought that perhaps the contents had escaped through a faulty valve. Lifting the cylinder, he realized that its weight had not changed. It was still full. Plunkett and an assistant sawed the cylinder in half and were shocked to find an extremely slippery white powder inside. What could this stuff be? Plunkett's fertile mind tackled the question and soon came up with an answer. The tetrafluoroethylene in the cylinder had reacted with itself, the small molecules joining together to make a giant molecule, or a polymer.

The substance had fascinating properties: it would not burn, it was resistant to corrosive materials, it was impervious to mold and bacteria, and it was more slippery than wet ice on wet ice. Because it was made from tetrafluoroethylene, DuPont christened the substance "Teflon." It signaled the dawn of a new era, but the world would not know about it for another eight years. Teflon would be classified as top secret by the U.S. government.

That's because it turned out to be the very substance that the boys at Oak Ridge were looking for. Teflon fittings could stand up to the corrosive gases involved in producing the U-235, the uranium species the military so desperately wanted. The Manhattan Project, in a sense, was lubricated with Teflon. After the war Teflon made its first public appearance, and people went wild. Everyone wanted to try the newfangled pots and pans lined with the slippery stuff that promised to put an end to scouring. Early versions of Teflon cookware did have one problem, though: the coating would flake off and end up in the food. Yet this was not a health concern, but an esthetic one, because Teflon is inert and passes through the body undigested. This particular problem could have been predicted; after all, if Teflon is the ultimate nonstick material, how could manufacturers get it to adhere to the cookware surface? Here's the modern secret.

First the metal base is washed with a solvent to remove any dirt, which could interfere with the binding process. The item is then blasted with minute particles of aluminum oxide, which create tiny pockmarks, greatly increasing the surface area. Then a primer, a mixture of Teflon and a binder, is sprayed on. The binder is a plastic, usually an epoxy, that sticks to the surface and provides a matrix in which the Teflon particles are embedded. This is heat cured, and then more layers of Teflon can be applied, because Teflon does stick to itself. Modern

Teflon cookware, like Silverstone, is a far cry from the early versions.

Today Teflon has numerous uses. From this marvelous substance we are able to manufacture artificial veins, arteries, ligaments, and hip joints, as well as spacesuits, Velcro, cable insulation, and Gore-Tex breathable fabric. In the 1970s Bill and Bob Gore had the idea of heating Teflon and stretching it into a fabric. They didn't have much luck — the heated Teflon always snapped. Then one day, in a fit of anger, Bob Gore took a hot piece of Teflon and gave it a violent pull. Now it stretched. Sometimes we can orient giant molecules, of which Teflon is one, by physical manipulation. Gore's rapid tug caused the molecules to line up, much the way a wad of spaghetti will straighten out if we pull on it. The uncoiled molecules in Gore's Teflon intertwined to make cables that were very strong and resilient. They refined the technique for making Teflon fabric, and Gore-Tex was born. This wondrous material has pores large enough to allow water vapor molecules to pass through but too small for the passage of water droplets. Hence dryness, no sweat.

Teflon's slippery quality has even led to the invention of coated bullets that can slice through bulletproof vests — not one of the substance's most endearing applications. In combination with nylon, Teflon is used in Stainmaster carpets. Dirt slides right off, just as it slides off Teflon politicians. Even Big Bird's feathers are protected by a Teflon coating; but, as far as smaller birds go, Teflon can represent a hazard. Overheated Teflon-lined pans give off vapors that can be toxic to pet birds. But Teflon is great for us. The promise of the product has indeed been fulfilled. We can prepare a great meal and clean up afterwards with relative ease. And if any food fragments become lodged between our teeth, we just have to reach for the dental floss. Give Glide floss a try — it slides smoothly through the narrow crevices and won't fray or shred. It's made of Teflon, of course.

Asbestos Panic

Charlemagne, Holy Roman Emperor from 800 to 814 A.D., was bent on gaining authority over the barbarian tribes he regarded as his enemies. When weapons didn't work, he resorted to magic. He invited tribal leaders to a dinner at which he planned to persuade them to convert to Christianity. After they had eaten, Charlemagne snatched up the ceremonial cloth from the dining table and hurled it into the fire. The fire went out, Charlemagne retrieved the cloth with a flourish, and he showed his guests that it was not only undamaged, but also clean. Surely this was magic.

Not really. Charlemagne was aware of the miraculous properties of a mineral fiber composed of various versions of magnesium silicate. In fact, the ancient Romans had discovered that this naturally occurring material was the ideal wick for the oil lamps of the vestal virgins, the watchers of the perpetual flame on the goddess Vesta's altar, because it did not burn away. We know this fiber as asbestos.

Asbestos will neither burn nor be corroded by acids or bases. That's why it is part of the fabric of modern society. It may be found in a host of items, from firemen's protective apparel and insulating material to floor tiles and cigarette filters; and it was long used to reinforce cement products, such as water pipes. Asbestos was even used in the construction of the Berlin Wall. But today asbestos seems to be going the way of the Wall — it's crumbling, both metaphorically and physically. And therein lies a problem: when asbestos crumbles, it releases fibers that can damage the lungs. In fact, people who scavenged souvenir chunks of the Berlin Wall have been advised to keep them in sealed containers.

Most of us don't have to worry about crumbling communist relics, but we may have to deal with other forms of environmental asbestos contamination, in our workplaces, our homes, and our schools. What's the real risk, and how much worrying should we do? The matter certainly warrants some thought, and in some cases action, but it does not warrant panic. Let's see why.

First let me tell you what you will not find in this discussion. Statements of certainty. Nobody can guarantee that a child will never be affected by airborne asbestos fibers in school, especially given that the effects may not show up for thirty years or more. The best we can do is make a careful, educated estimate of the risk based upon the available scientific data. Second, you should know that there is much controversy, even among experts, over how to handle the asbestos problem in a nonoccupational setting. This, in itself, is an indication that the risk is small, otherwise we would not be debating the issue. We don't debate whether smoking cigarettes or driving without a seat belt are dangerous activities.

We don't debate whether asbestos is a carcinogen, either. It is. There is absolutely no doubt that asbestos miners, asbestos insulation installers, and asbestos textile workers show a higher incidence of lung cancer and a higher incidence of mesothelioma, a fatal cancer of the membrane that lines the lungs. But the incidence of these cancers is related to the length of exposure, the concentration of asbestos fibers in the inhaled air, and the type of asbestos involved. Furthermore, the risk of lung cancer due to asbestos is clearly linked to smoking: the incidence of the disease is ten times greater among smokers.

Although we speak of asbestos as if it were a single substance, it actually isn't. There are several forms of the mineral, but they can all be classified as chrysotile or amphibole, depending on the size and shape of the fibers. Chrysotile fibers are fine, curly, and flexible, while amphibole fibers are straight and brittle. Their effects on the lungs are quite different. Chrysotile fibers can be cleared from the lungs far more readily, and they cause less damage. The amphibole fibers are the real nasties.

In occupational environments workers are exposed to several forms of asbestos, but in the general environment ninety-nine percent of airborne fibers are chrysotiles. This makes it hard for us to extrapolate risk from an occupational to a nonoccupational setting, because we may not be dealing with the same material. In fact, a landmark study carried out by researchers at McGill University, Laval University, and Health Canada examined disease rates among women living in asbestos mining areas but not directly exposed to asbestos and found that the commonly used risk-assessment models, as developed by the American Environmental Protection Agency, greatly overestimated the incidence of lung problems. Still, the prudent approach would be to assume that all of the fibers present are dangerous. Surveys have shown that an increased risk of lung damage becomes apparent at a concentration of fifty to one hundred fibers per cubic centimeter (cc) of air. To ensure at least a one-hundredfold safety factor, occupational exposure limits are usually set in the range of 0.2 to 0.5 fibers per cc.

How does this compare to levels that have been measured in schools where asbestos-based materials have been used? Typical levels found in such schools are in the range of 0.0005 fibers per cc, and the highest readings ever recorded have been approximately 0.01 fibers per cc. Epidemiologists have calculated that if schoolchildren were exposed to such levels for five years, there would eventually be one lung cancer death attributable to asbestos for every one hundred million of them. This does not sound very risky, so why all the fuss? Because there is always the theoretical chance that levels will increase significantly if asbestos insulation becomes friable — in other words, if it disintegrates and releases airborne fibers. This can happen due to aging, vandalism, or water damage.

If asbestos becomes visibly damaged, then it needs attention. Removing it is not necessarily the answer, since improper removal of asbestos actually increases the concentration of fibers in the air, and proper removal is expensive. (Do we really want this job to go to the lowest bidder?) We may use various coatings to secure the disintegrating asbestos; the W.R. Grace company has come up with a product that is applied as a foam and chemically converts asbestos into harmless oxides of silicon and magnesium while retaining fire-retardant properties. Alternatively, we may install false ceilings or drywall. Another option is to moisten flaking pipe insulation to reduce dust and then wrap it with duct tape. If the asbestos is undamaged, our best bet is to leave it alone. It will not jump out and attack children.

Asbestos may, however, attack carpenters, electricians, plumbers, and school custodians who inadvertently disturb it. These people have to exercise due caution and use protective equipment whenever they think they may be dealing with asbestos. As for the rest of us, I don't think asbestos is much of an issue. Would I like to work in an asbestos mine? No way. Would I want to be drilling through asbestos panels? Uh, uh. Would I hesitate to send my children to a school constructed with asbestos-containing materials? Nope. But I would like to know that an expert who has up-to-date knowledge of asbestos inspects the building periodically. And if that expert should recommend action, I would certainly not insist that it begin the next day.

Let me leave the final word to James Peto and Richard Doll, two of the most respected epidemiologists in the world. Both suggest that a person who spends an hour a day in a room with a smoker is one hundred times more likely to contract lung cancer than a person who spends twenty years in a building that contains asbestos. So, take a deep breath and relax — unless you're near a smoker.

Murder by Thallium

Graham Young didn't kill people out of malice. The youthful chemistry buff just enjoyed investigating poisons, even going as far as using his stepmother's body as a laboratory. He almost killed her. Eventually, Young made other relatives, friends, and coworkers his guinea pigs. His one-man chemical reign of terror finally came to an end when the cocky fellow suggested to befuddled investigators that the symptoms of all those sick and dying people in his life seemed to indicate thallium poisoning. Indeed they did, and the trail led straight to him. In 1972 Young was sentenced to life imprisonment.

Most people don't know that thallium is one of the chemical elements. In fact, it is among the most toxic. Its name derives from the Greek word *thallos*, meaning a "budding shoot or twig," because the metal emits a bright green color when heated to a high temperature. The element was discovered in 1861 by the English chemist William Crookes, who is better known for contributing to our knowledge about electrons and dabbling in the supernatural. Thallium compounds occur naturally in minerals, but they have no real commercial importance except for their occasional use as a catalyst in some chemical reactions. The most interesting thallium application involves the use of an isotope, thallium-201, in diagnosing heart disease. The radioactive isotope is injected into the patient, and it then binds to heart muscle, but only if the tissue receives an adequate supply of blood. If there is inadequate blood flow to an area of the heart, thallium will not bind. Since thallium-201 gives off gamma rays, which readily go through tissue, an outside detector can reveal a lack of thallium binding and a probable arterial blockage. No need to worry about toxicity here. Very small amounts of thallium are needed, and the radioactivity quickly disappears.

Until the middle of the twentieth century, though, thallium sulfate was widely used as a commercial rat poison. Zelio Paste, sold in a tube, hastened the demise of the revolting creatures. Unfortunately, rats weren't the only creatures that succumbed to the toxin. Since 1935 almost eight hundred people have suffered thallium poisoning and nearly fifty have died. Most of these poisonings resulted from accidents or improper use of the chemical. Perhaps the most fascinating case of thallium poisoning involved a nineteen-month-old baby who was brought to London's Hammersmith Hospital from the Persian Gulf state of Qatar in 1976. She had suffered a seizure of sorts, which appeared to have left her permanently disabled. Her movements were jerky and uncoordinated; her speech was slurred. A physical examination and blood tests revealed nothing. The child's doctors were stumped, and they could only stand by and watch her slide towards death.

Then a remarkable thing happened. A nurse on the intensive care ward where the child had been placed for observation was an avid Agatha Christie fan. She had just been reading one of Dame Agatha's classic mystery novels, *The Pale Horse*. The plot revolves around an organization that eliminates "undesirables" for a fee by sneaking poison into their food, their drink, or even their toothpaste. It occurred to Marsha Maitland, the Hammersmith nurse, that the symptoms of Christie's victims were remarkably like those of the sick child. When she pulled on the baby's hair and noted how easily it came out, she became convinced. Just like in the mystery story, this was thallium poisoning.

Maitland told the doctors about her suspicions and they ordered a special urine test, which revealed the presence of thallium. They finally determined that the child had accidentally consumed some thallium sulfate, which in Qatar was widely used to combat cockroaches and rodents in drains and septic tanks. Thallium poisoning is hardly a subject that comes up often in medical school, so the Hammersmith doctors scrambled for information on how to treat their small patient. Surely Scotland Yard knew something about exotic poisons. It did, thanks to the case of Graham Young. The doctor who had looked after Young's victims had done a comprehensive search of the literature on thallium poisoning. He had learned that thallium is chemically and biologically similar to potassium and can displace it in certain biochemical reactions. Potassium is important in activating a number of enzymes that cannot function in its absence. Hair follicles and cells of the nervous system are the first to be affected. Loss of hair, lack of muscle coordination, pain in the abdomen and the extremities, as well as diarrhea and nausea are classic signs of thallium poisoning.

The treatment of thallium poisoning is rather ingenious. It takes three forms: diuretics eliminate thallium through increased urine production; Prussian blue dye traps thallium excreted into the gut, preventing its reabsorption; and potassium chloride supplements displace the thallium that has already been absorbed into the tissues. These measures allowed the unfortunate child from Qatar to make a complete recovery.

In 1989 an even more mysterious case of thallium poisoning occurred in what was then the Soviet Union. In two separate towns, one in the Ukraine and the other in Estonia, dozens of children began to exhibit the symptoms of thallium poisoning. Doctors confirmed the presence of thallium in their bodies.

But how did it get there? The likelihood is that someone with a bit of chemical knowledge, but not enough, added thallium compounds to local gasoline supplies to improve its performance. These compounds are similar in chemical behavior to tetraethyl lead, which increases the octane rating of gasoline. Thallium was likely released in car exhaust and eventually found its way into the children's bodies.

George James Trepal of Florida knew all about the chemistry of thallium. He had used it as a catalyst in the clandestine laboratory where authorities nabbed him as he was preparing methamphetamine, an illegal substance. Trepal spent two years in jail for that crime, but he failed to learn that crime doesn't pay. In 1989 he became annoyed with his neighbor because she played her music too loud and let her dogs bark. So he spiked her Coca-Cola with thallium. Trepal, a member of Mensa, an organization for people with high IQs, planned to commit the perfect crime.

He knew that thallium compounds had no smell or taste and that the symptoms of poisoning could easily be mistaken for a variety of other ailments. But he wasn't quite as smart as he thought — Trepal neglected to get rid of the evidence. Police found a vial containing remnants of thallium nitrate in his home. George James Trepal now sits on death row while his lawyers attempt to cast doubt on the chemical evidence. And he's not the only thallium murderer to have received the death sentence. In Pennsylvania, Joann Curley awaits the same fate.

The prosecution argued that Curley had planned to poison her husband, Robert, in cold blood for his life insurance money. She seized the opportunity when Robert Curley took a job that involved remodeling the labs in the chemistry building at Wilkes University. Joann began to add thallium to the ice tea that Robert took to work each day in a thermos, thinking that even if the authorities eventually made a diagnosis of thallium poisoning, then they would link it to Robert's exposure to thallium compounds in the lab. She miscalculated. Toxicologists quickly determined that laboratory exposure could not account for Robert's condition. Joann's fate was sealed when she became impatient to finish off her husband and took him some ice tea when she went to see him in the hospital. The suspicious authorities had the tea analyzed, and Joann Curley was arrested. The courts have yet to decide the fates of Trepal and Curley, but Graham Young won't be experimenting with thallium anymore. He died in prison of a heart attack at the age of forty-five. Poisoning is a stressful business.

Danger down the Drain

Do not try this at home. I'm serious. The consequences can be disastrous. Those who possess some chemical expertise, however, can safely launch a hydrogen rocket, and the result will be spectacular. In fact, it's one of my favorite demonstrations. I generate hydrogen gas by reacting sodium hydroxide with aluminum, and then I store the gas in a plastic soft-drink bottle. Taking all the necessary precautions, I then ignite the gas, producing a reverberating bang and a sensational lift-off. But remember this warning: unless you know exactly what you're doing when conducting this demonstration, you may end up drenched with a horribly caustic sodium hydroxide solution.

Okay — I've scared you enough. You're now well aware of the danger involved in playing with homemade hydrogen rockets. So what's the point? My aim here is to show you that you may be exposing yourself to the same type of hazard as you perform the seemingly innocuous task of cleaning your drains. Drains get clogged. That's a fact of life. Hair, soap scum, and grease are the usual culprits, but most plumbers can regale us with stories of the strange things — ranging from dead rats to live pythons — that they have extricated from the nether regions. That's right — live pythons. Singapore shotput champion Fok Choy was sitting on a toilet seat when a python bit him on a very sensitive part of his anatomy. The creature had snaked its way up the drainpipe. I guess that in Singapore it pays to look into the bowl before sitting down.

Given that all kinds of stuff — except, perhaps, for the kitchen sink itself — can be swallowed by a drain, it isn't surprising that chemical drain cleaners are hot-selling items. The most common type of product is a mix of flakes or pellets of sodium hydroxide, commonly known as lye, and little slivers of metallic aluminum. Sodium hydroxide breaks down fats in one of the oldest known chemical reactions, saponification. Eons ago our ancestors noted that when animal fats were mixed with ashes from a fire, a novel substance, which foamed when wetted, was formed. This, of course, was soap. Ashes contain a variety of basic or alkaline substances that create soaps from fats. When we dump sodium hydroxide down a blocked drain, it reacts with grease to make a water-soluble soap. The drain-cleaning effect is boosted by the fact that the dissolution of sodium hydroxide is an extremely exothermic process and causes fat to melt, allowing it to be flushed away more easily. Sodium hydroxide also degrades proteins, like hair, breaking them down into water-soluble amino acids.

But why do manufacturers add bits of aluminum to drain cleaners? When immersed in water, the aluminum reacts with sodium hydroxide to generate hydrogen gas, just as it does in my rocket. This generates more heat and supposedly helps to dislodge deposits through effervescence. "Supposedly" is a key word here. These chemical drain cleaners don't work all that well for the simple reason that they don't usually reach the site of the blockage. Generally, the problem does not lie in the U trap under the sink or toilet but further down the line, where the pipe makes a sharp downward turn. By the time the drain cleaner gets to this area, it has been diluted to the point of ineffectiveness. But if the blockage is, indeed, in the U trap, another problem can arise. The sodium hydroxide gets trapped, a fantastic amount of heat is generated, and a toxic mix of steam and lye comes shooting back up from the drain. Never peer down the drain to see what's happening. If your drainpipe is still blocked after a dose of sodium hydroxide mixed with aluminum, then you may become frustrated and attempt to unclog it with a different chemical. Concentrated sulfuric acid is also sold as a drain cleaner, but chasing sodium hydroxide with such a concoction is a bad idea. A very bad idea. The two types of product react to neutralize each other, but in the process they may produce enough heat to destroy your pipes.

Your misadventure will likely end with an emergency call to the plumber. Most plumbers have a love-hate relationship with chemical drain cleaners. They love them because when do-it-yourselfers damage their pipes there's money to be made; they hate them because working with pipes that are loaded with caustic reagents is no picnic. The solution most plumbers will opt for is poking an electric fish wire, up to seventy-five feet in length, down the drain and switching the power on. The cable will rotate frantically, macerating any blockages.

Given that most of us are not equipped with such sophisticated weaponry and that chemical drain cleaners are not ideal, what do we do? First and foremost, practice preventive maintenance. Make sure that fat does not go down the drain. Wipe greasy plates with soiled napkins; newspaper is great for absorbing oil. If you have a garbarator, then use plenty of cold water to flush away the remains; carrot peelings and corn husks, particularly, can cause obstructions. To keep drains flowing smoothly, mix one cup of bicarbonate, one cup of salt, and a quarter of a cup of cream of tartar in a container, then, once a week, pour about three spoonfuls down the drain followed by a kettleful of boiling water. The bicarbonate and cream of tartar react to generate carbon dioxide gas, which dislodges small deposits, and the salt ensures that the solution is dense enough to flush down the pipe.

If, in spite of all this, your drain gets clogged, try opening it up with a plunger first. Should that fail, try dissolving half a pound of washing soda in a gallon of boiling water and pouring this down the drain. Enzyme preparations, advertised as "natural and nontoxic" are usually ineffective. You can experiment with Drano, or another product of its ilk, but make sure you follow the directions carefully and do not overuse the stuff. And make sure, as well, to tell the plumber — whom you will probably end up calling — to charge a little extra as compensation for the fact that you've made his life more difficult by filling the drain with nasty things.

Obviously, clogged drains are not much fun. But free-flowing ones can be. They afford fantastic opportunities for scientific conjecture and experimentation. For example, does water really spiral down the drain counterclockwise in the Northern Hemisphere and clockwise in the Southern? Certainly, hurricanes and tornadoes spin in opposite directions above and below the equator. This is due to the so-called Coriolus effect, which is based on the fact that the earth spins in an easterly direction at a speed of about a thousand miles per hour at the equator. As we move north or south the speed decreases, since less distance has to be covered in the same time. Now, imagine that there is a gigantic bathtub extending from the equator to the North Pole. The water near the equator will travel east much faster than the water near the North Pole. If we pull the plug, the water will spin down the drain counterclockwise due to the momentum imparted by the eastward-traveling southern water. The opposite would be true in the Southern Hemisphere.

While that holds true for a giant tub, there's no point in studying your own bathtub drain. The Coriolus effect for such a small amount of water is too tiny to be observed; and, in any case, the direction in which the water drains is determined by factors such as the slope of the tub or shape of the drain. But what would happen right on the equator? I'm told that a certain hotel in Kenya is geographically situated so that the water in its bathtubs drains straight down. I'm skeptical, but this would be an interesting phenomenon to investigate. Anyone interested? I can tell you that it's a safer science project than generating hydrogen gas from drain cleaner.

The Poop on Methane

The newspaper headline grabbed everyone's attention: "Holy Blam, Batman! Bat Dung Brews Methane; Park Building Explodes!" The story seemed to support its sensational title. It intrigued readers with its description of a remarkable event that occurred on June 12, 1993 in a Michigan state park. An explosion had rocked the area and reduced an abandoned ranger station to smithereens. At first police investigators could find no cause for the explosion. There was a propane furnace in the ranger station, but it had long since been disconnected and the gas had been shut off at the outside propane tank, which was still intact. No one held any insurance on the dilapidated building, so foul play was ruled out. But then one of the police officers noticed that the floor inside the building was soiled with black gunk that had apparently dripped down from the attic. He decided to investigate. As he opened the attic door, he was startled by the frantic beating of hundreds of wings. Bats had invaded the abandoned house and roosted in the attic. The quantity of slimy stuff on the floor was evidence that they had been there a long time: it was bat guano.

The police officer knew that organic waste, including bat poop, could produce a combustible gas. A possible explanation for the blast began to form in his mind. He hit the books and learned that manure contains a bacteria that converts the various organic molecules that have survived digestion into methane gas. We normally refer to this as "natural gas." Methane burns, and, under the right conditions, it will explode. A zoologist confirmed for the officer that guano did, indeed, have the potential to produce methane, and he added that methane was heavier than air and had therefore probably settled in the basement. The fire marshal's report concluded that the explosion likely occurred when the methane was ignited by a spark from a sump pump that was still functioning in the basement. It was a neat, interesting, and sensation-provoking report — but there was a problem. It was wrong.

The investigators should have consulted a chemist instead of a zoologist. They would have learned that methane is lighter, not heavier, than air. The methane would have escaped through the openings in the roof and could not have accumulated in the basement. Had they consulted a microbiologist, they would also have discovered that the bacteria that produce methane from waste are anaerobic, meaning that they can only survive in the absence of air. Bat droppings are quite porous, so the bacteria they contain would be exposed to the air, and no appreciable amount of methane would be produced. If such conditions alone did produce a lot of methane, then we would hear all kinds of reports of visitors to bat-infested caves blowing themselves up by striking matches.

What, then, triggered the blast at the ranger station? In light of the new scientific revelations, the investigation was reopened.

The culprit turned out to be manure, all right, but of a decidedly human variety. The building's toilets connected to a septic tank, with the usual U-shaped trap built into the pipes. These traps are normally filled with water, which prevents any gas from backing up into the toilet bowl. Since septic tanks create the anaerobic conditions under which flammable methane can be generated, this is an essential safety measure. The toilets had not been used since the station was abandoned, and the water had evaporated from the plumbing system. Methane was now free to back up into the basement, and it had apparently done so. As the original investigator had proposed, a spark from the sump pump had probably set off the explosion. The bats had been falsely accused.

Such methane explosions are a real danger wherever organic waste decomposes in an environment with restricted air flow. Landfills are ideal for the production of methane. This can be a blessing or a curse. If we recover the gas, which we now do at many landfills, we can burn it as a source of energy. But if we don't allow the gas to vent and let it build up underground, catastrophe may result. This is not just a theoretical concern. A bungalow in England blew up in this fashion, injuring three people and forcing the evacuation of an entire neighborhood. The house was next to a landfill composed of local trash. On the fateful day the weather changed dramatically as a low pressure front swept in. Due to the pressure differential, the methane gas that had formed in the landfill began to flow towards the surface, and it accumulated under the floor of the house. The residents of the house had turned the heat on, and the rising hot air created convection currents that sucked the methane into the basement. The gas was probably ignited by the furnace's pilot light. Boom!

For those of you who are disappointed by the discrediting of the more exotic bat-poop theory, let me offer the following

item, which comes to us from the reliable Reuters news service and originates in Granada, Spain: "Nearly five hundred pigs were blown up or burned to death when a short circuit ignited methane gas produced by their own dung. Twenty of the animals escaped by battering down one of the doors of their shed." Pigs produce a lot more dung than bats, and it is conceivable that an appreciable amount of methane built up in the confined space they inhabited. And if a combustible gas in a confined space is ignited, there will be an explosion.

One of the most curious demonstrations of this effect was inadvertently carried out by a farmer in Cluj, Romania. Here people traditionally slaughter pigs at holiday time, and many will inflate the carcass with a pump or the exhaust from a vacuum cleaner to make the skin taut before removing the animal's bristles with burning straw. Our farmer's vacuum cleaner had broken down, so the resourceful man decided to use bottled gas instead. Not a good decision. When he began to singe off the hairs, the pig exploded, striking and injuring the farmer. And they say pigs can't fly.

Mulish on Borax

One day a craftsman arrived at the court of Tiberius, emperor of Rome from 14 A.D. to 37 A.D. He had come to present the ruler with a beautiful transparent glass vase. Just as he handed it to Tiberius, the man let the vessel slip through his fingers. Everyone jumped, expecting a crash and flying shards of glass — Roman glass was notoriously fragile. But the glass did not break. Tiberius was very impressed and asked the glassmaker to tell him the secret of the unbreakable vase. The craftsman refused to divulge it and boasted that he alone could make this marvelous glass. This did not sit well with the ruthless emperor, who had the glassmaker put to death. The unfortunate artisan had probably stumbled upon a method of making borosilicate glass. Ordinary glass is made by heating together sand, limestone, and sodium oxide, a process that was well known to the Ancient Romans. If certain boron compounds are added to the mix, the glass becomes shatter resistant. This is the kind of glass we use today in laboratories.

Sodium borate, or borax, was used by Roman goldsmiths as a flux — a substance added to the metal to make it flow more easily when heated. We can only hypothesize that the glassmaker added some borax to his molten mix to improve its flow properties and noted that the resulting glass was much stronger. The Romans never did discover the secret, and it wasn't until 1916 that an American patent officially described shatterproof Pyrex glass, a product containing a high percentage of boric oxide. Pyrex can withstand extreme temperature changes without cracking, and it is ideal for ovenware.

Borax was a rarity in Ancient Rome, found only near a few Tuscan hot springs. But then, in the thirteenth century, Marco Polo returned from the Orient with samples of gunpowder, spaghetti, and white crystals of borax. The Italians quickly figured out that they could make spaghetti themselves, but they'd have to import the borax. They organized caravans, and soon Venetian goldsmiths were enjoying an ample supply of this superior welding flux. Europeans also learned that the Chinese used borax in pottery glazes and began trying it themselves for this purpose. Using borax to improve the flow qualities of molten glass was a natural extension of the process.

Yet the substance was still hard to come by. The age of borax did not begin until 1870, when huge natural deposits were found in the Nevada desert. It was there for the taking; all one needed was a shovel. Before long prospectors were crawling all over the desert in search of the richest deposits of the mineral. The borax rush was on. Since most people would have difficulty recognizing borax on sight, a simple identification test was needed. Prospectors discovered that they could pour sulfuric acid and alcohol on a piece of ore, ignite it, and watch for the green flame that would signify the presence of borax.

Sulfuric acid reacts with sodium borate to form boric acid, which colors the flame generated by the burning alcohol green. We still use the green-flame test to identify the element. In 1881 it was just such a flame that prompted grizzled Death Valley prospector Aaron Winters to let out a scream that brought his wife running: "She burns green, Rosie! We're rich!" Indeed, the find was a major one — Winters sold his claim for the staggering amount of twenty-thousand dollars to developer William T. Coleman. But there was a problem. How could they move huge amounts of borax out of the desert without a railroad link? The inventive Coleman came up with the idea of employing gigantic wagons drawn by a twenty-mule team. Fully loaded, these conveyances weighed thirty-two thousand pounds each, and their wheels were seven feet high.

The twenty-day roundtrip through the desert to Mojave, the nearest railroad junction, was treacherous. Temperatures

often exceeded 110°F. Nevertheless, in six years twenty million pounds of borax were moved out by the twenty-mule teams. This was more borax than the world needed, and Coleman eventually went bankrupt. At this point F.M. "Borax" Smith entered the picture and resuscitated the company with his imaginative ideas. He was going to make borax a household staple. He put out advertisements hyping the product as a complexion aid, a milk preservative, a cure for epilepsy and bunions, and an additive for bath water and the water used to wash carriages. The latter claim proved to be sound: borax binds minerals that can interfere with the cleaning action of soaps and detergents. Indeed, today we still use borax in laundry products. If you want to improve the appearance of your laundry and cut down on your detergent consumption, just add about a third of a cup of borax to each load. A quarter of a cup of borax dissolved in two cups of water is also a great spot remover for carpets.

Most of the world's supply of borax now comes from a gigantic open-pit mine near a California town appropriately named Boron. There are now myriad uses for the material that started its technological life as a gold flux. We use it to make glass for car headlights, enamel for stoves and refrigerators, ceramic tiles, antiseptics, bleach for unbleachables, weed killers, and fertilizers. Thin fibers made from elemental boron reinforce resins destined for aircraft and space-vehicle parts.

Boric acid, which is made from borax, can control cockroaches. Offer them flour and sugar bait spiced with boric acid, and you'll likely be serving them their last supper. Even if a cockroach chooses not to dine, the crystals of boric acid will stick to its exoskeleton and rub away the oily protective layer. The creature will dehydrate and die. An ant will suffer a similar fate. Finally, borax can even teach us some neat chemistry. Just combine half a tablespoon of it with half a cup of white glue, and stir. Almost miraculously, the sticky glue changes into a putty that can be stretched or rolled into a ball. Why does this happen? The glue is made of long molecules of polyvinyl acetate, which easily slip and slide past each other. Borax links these long molecules together, forming a semirigid three-dimensional network of molecules that is neither a liquid nor a solid. But it is a lot of fun. Tiberius would have loved it.

The Silly Side of Flubber

I went to see *Flubber*. I had to. How could I not go and see a movie about a chemistry professor? I was worried, however, that it might be a painful experience, since moviemakers seem bent on portraying chemists as absentminded, bumbling nincompoops. My fears were well founded, but this professor, played by Robin Williams, wasn't absentminded — he was downright demented. *Flubber* is a remake of the 1961 picture *The Absent-Minded Professor*. In the original, the chemist character, played by Fred MacMurray, was actually charming, and the story was rather engaging, especially for anyone familiar with the energy that had been devoted to rubber research in the previous two decades.

According to Disney insiders, the idea for "flying rubber" evolved from some fascinating experiments that Dow Corning and General Electric had carried out in their research laboratories during the 1940s and 1950s. The goal was to develop some sort of synthetic rubber. During World War II the United States feared being cut off by the Japanese from its natural rubber supplies in the Orient. Rubber was tremendously important to the war effort — it was needed to make equipment ranging from tires and gaskets to boots and gas masks. A replacement for natural rubber had to be found. Scientists at G.E. and Dow Corning turned their attention to a class of substances that had recently been discovered by Frederick Stanley Kipping at the University of Nottingham. Kipping had become interested in the chemistry of silicon because of the similarity of the chemical behavior of this element to that of carbon. He knew that human life was essentially based on the ability of carbon atoms to combine with each other; the most important biomolecules, like proteins, carbohydrates, fats, and DNA, all have basic skeletons composed of carbon.

The British scientist wondered whether he could produce similar substances based on a silicon framework. His efforts inadvertently planted the seeds of an idea that has since blossomed in the mind of many science fiction writers: an alien world may exist where everything is based on the chemistry of silicon instead of carbon. Kipping, however, by his own admission, did not think that he had come up with anything dramatic. He produced a variety of oils and gunks, which he recognized as being composed of long molecules with a backbone of silicon and oxygen atoms. These apparently useless substances he named "silicones."

But American chemists searching for a rubber substitute did not miss the fact that some of Kipping's silicones had a rubbery texture. At General Electric, James Wright was in charge of a project to develop a synthetic rubber insulation for new electric submarine motors. Silicones were water resistant, had great insulating properties, and would be ideal if a rubber-like resiliency could be achieved. Wright began to tinker with the silicone molecules, hoping to achieve flexibility by crosslinking the long molecular chains. The chains could still stretch, he theorized, but they should recoil to their original shape when the applied force was relaxed. He thought that by incorporating boron atoms into the framework he could create points at which to attach adjacent polymer chains. The theory proved to be correct. Wright's addition of boric acid to silicone oil produced a marvelous new material. The pliable mass could be rolled up into a ball that bounced twenty-five percent better than a comparable rubber ball. Here was the desired property. Unfortunately, the substance had some undesirable properties, as well. When you hit it with a hammer, the stuff shattered; when you pulled it quickly, it broke apart. A fascinating but useless substance.

In 1945 G.E. sent thousands of samples to engineers around the world, hoping that they could come up with some use for the new material, but the company was not rewarded with any brilliant insights. Wright went on to other work, but he kept a sample of the putty to amuse his friends with — everyone was charmed by the substance that could be made to bounce crazily when shaped into a ball, yet spread itself into a blob when left on a table. Wright sometimes even took a sample to cocktail parties to liven things up. In 1949 a momentous event occurred at one of these parties. Wright, as he had so often done before, began to bounce his invention. As usual, a crowd gathered. This time it included toy store owner Ruth Fallgatter and a man she'd hired to prepare a catalogue, Paul Hodgson. Both were immediately taken with the bizarre properties of the material, and when Wright demonstrated how it could even lift colored images off paper, Fallgatter and Hodgson were sold.

After getting permission from General Electric to market the stuff, the pair listed it in their catalogue under the name Nutty Putty. Although the putty sold well, for some reason Fallgatter lost interest. But Hodgson didn't. This clever man bought a ton of the silicone from General Electric for $147, and he packaged it in one-ounce, egg-shaped containers. His investment turned out to be one of the shrewdest ever made.

When Silly Putty, as Hodgson had renamed the product, was mentioned in *The New Yorker* magazine, Hodgson received orders for three quarters of a million eggs in three days. In the first year sales figures reached an unheard-of six million dollars. By the time Hodgson died he was worth 140 million dollars. Not bad for a "useless material." Toy manufacturers Binney and Smith now own the rights to the name Silly Putty, but other companies sell the same material under a host of names. Interestingly, nobody has yet found any other important commercial application for the strange substance. Some have used it to level chairs or exercise injured hands; others find it handy for removing lint from clothes, dirt from keyboards, or stains from artificial limbs. There are therapists who believe that playing with Silly Putty has a calming effect and can relieve stress. But perhaps the most intriguing suggestion I've ever heard came from a gentleman who liked to hurl a wad of the putty at the newspaper stock market listings and invest in whatever stock it lifted off the page. I don't know if the man ever made any market gains, but Binney and Smith awarded him a fourteen-carat-gold blob of Silly Putty for his ingenuity.

Scientists have, however, developed other silicones that boast a large variety of practical uses. We put them to work as lubricants, antifoaming agents, water repellants, car waxes, breast implants, hand cream and shampoo ingredients, sealants, gaskets, sterilizable baby bottle nipples, oven door seals, and heart valves. Kipping and Wright would be amazed. Then there is Flubber. Disney was so enthralled with the amazing properties of Silly Putty and the accidental nature of its discovery that the studio decided to create its own bouncy story. And what did they use in the 1961 movie to play the role of the Flubber? I'll let you in on a secret: it was a concoction made from salt water taffy, yeast, polyurethane foam, cracked rice, and molasses.

The new version of Flubber did not result from such chemical ingenuity. Disney employed computer graphics to fabricate the green ooze that sends Robin Williams into his annoying hyperactive fits. Williams got to costar with a substance that served as a robot helper, spoke with an alluring female voice, and had human thought processes. Still, the only thing *Flubber* did for me was to rekindle my interest in *The Absent-Minded Professor*. I decided to rent the video. But I never did. I forgot.

From Jewelry to Jupiter

A friend and I were talking, and the conversation turned to the subject of online auctions and all the neat things you can buy. I mentioned to her that I had been looking for a vividly colored Bakelite radio, but the ones I liked were quite pricey. My friend, the owner of some Bakelite jewelry, agreed that it was an attractive material but added that it was "just like plastic." Well, Bakelite isn't *like* plastic. It *is* a plastic. And, for me, that's no drawback — Bakelite is an astonishing material.

We wouldn't even have it if George Eastman had not invited

Leo Baekeland to meet with him one fateful day back in 1898. "Come in, come in, Dr. Baekeland," the founder of the Kodak empire said to his Belgian guest. "I think you will find this chair very comfortable." The meeting went well, and as a result we now have plywood, particle board, heat-resistant pot handles, heat shields for spacecraft, beautiful old radios, and a fascinating story.

Leo Baekeland was educated at the University of Ghent, in Belgium, where he fell in love with chemistry, as well as with his professor's daughter. This did not make Professor Theodore Swarts particularly happy, because he thought that young Leo should be more concerned with the chemistry occurring in the lab than the chemistry occurring in the Swarts family parlor. Swarts was further annoyed to learn that his brilliant student was also devoting time to another interest. Leo wanted to make money. He was constantly hatching schemes to develop a better process for making photographic plates. At the time these plates were coated with silver compounds that turned black when exposed to light — a lot of light. Without bright sunshine or a magnesium flash, the photographer could take no pictures.

Baekeland had long been interested in the chemistry of silver compounds and in improving the photographic process. America, he thought, would offer him the best opportunity to pursue his interests. After spending some time in the chemistry department of New York's Columbia University and then working for a manufacturer of photographic equipment, Baekeland decided to go private. He dedicated himself to finding a better way to make photographic paper. Within two years he had done exactly that, and Velox paper took America by storm.

When Baekeland realized that he could not cope with large-scale manufacturing, he approached Eastman with the idea of selling him the process. He'd planned to ask for fifty thousand dollars, a large amount of money in those days, but he was ready to settle for half of that. Baekeland never got a chance to exercise his bargaining abilities. As soon as Baekeland sat down in the comfortable chair to which Eastman had directed him, Eastman made an offer: "Dr. Baekeland, how would $750,000 sound for Velox?" A deal was struck, and Baekeland became a wealthy man. In fact, he became so wealthy that he would never have to worry about money again, and he was free to devote his time to the chemical problems that intrigued him.

The female Indian lac bug, *Lacifer lacca*, had captured Baekeland's interest. This little creature was the source of all the world's shellac. It sucked tree sap and converted it into a resin. The resin was scraped off the trees and purified into a substance that, when applied to wood, provided a nice shiny finish. But now, within the burgeoning field of electricity, shellac had found a brand new use: it was an ideal insulating material. Unfortunately, it took fifteen thousand insects about six months to produce a pound of shellac. What the world needed was a cheap source of synthetic shellac, and Baekeland was going to find it.

Baekeland was not some would-be inventor but an accomplished chemist, familiar with the scientific literature. He knew that thirty years earlier Adolph von Baeyer had mixed together phenol, a chemical used as a disinfectant, with formaldehyde, a common preservative. The mixture had sizzled and foamed and produced a black tarry mess that had frustrated Baeyer because it would not dissolve in any known solvent. He couldn't even melt the stuff off his equipment. Baekeland also knew that Werner Kleeberg, in Germany, had recently developed an interest in this reaction because he had wanted to improve upon a new product called "Galalith." Fellow German Adolph Spitteler had developed the product with a little help from a clumsy cat. His pet had knocked a bottle of formaldehyde into its milk saucer. Spitteler, a chemist, noted that the milk turned into a tough, plastic-like material, and soon German schoolchildren were writing on washable boards that had been coated with the substance. Galalith inspired Kleeberg to have another look at the phenol-formaldehyde reaction to see if he could come up with a competitive product, but, like von Baeyer before him, he was unsuccessful in taming the reaction.

In 1902 Baekeland decided that he would have a go at it himself. It took him five years, but he did produce a synthetic shellac. He also drove his neighbors crazy by creating formaldehyde and phenol fumes, which emanated from his garage. The key to his success turned out to be running the reaction at a high temperature and under high pressure in a device that Baekeland called a "Bakelizer." The thick, amber-colored liquid he produced in this way was a great substitute for furniture shellac, but it was too brittle to serve as an insulating material. Yet Baekeland solved this problem, as well. When he added fillers such as cellulose fibers or wood flour to his mix, the final product became tougher than nails. It could even be colored. He modestly named the invention Bakelite.

By 1924 Bakelite had become so popular that it was featured on the cover of *Time* magazine as a substance that "will not burn and will not melt." Bakelite jewelry, telephones, pens, radios, car parts, airplane propellers, ashtrays, billiard balls, and cameras were everywhere. Just about the only item that did not become popular was the Bakelite coffin. People would wear, cook in, and eat off plastic, but they refused to be buried in it. Bakelite's success stimulated research to improve the material even further, especially after scientists pointed out that in the manufacture of Bakelite the small phenol and formaldehyde molecules had joined together to make a giant three-dimensional lattice. In other words, Bakelite was the world's first synthetic giant molecule, the first synthetic polymer.

Starting with this idea, chemists subjected other similar small molecules to the same conditions, and soon other plastics, like Melamine and Formica, appeared.

Over the years modern chemistry has given us a huge array of plastics, but Bakelite, the world's first synthetic plastic, is still widely used. It is the adhesive that bonds plywood and particle board, the mainstays of modern construction. Its heat-resisting and insulating properties make it ideal for pot handles. The 1995 Jupiter space probe had a heat shield made of Bakelite. But, alas, we no longer make telephones of this historic material. You can't get nearly as much satisfaction from slamming down a light, modern, polystyrene telephone receiver as you can from banging a good old black Bakelite one. And then there is Bakelite's secret use: making fake amber. How fascinating that is — here is a synthetic plastic developed as a substitute for a natural resin that now tries to mimic natural resin.

Perhaps now you understand my fascination with plastics. But I know that not everyone appreciates them as much as I do. When my wife and I went out to purchase a chandelier recently, we found one that we liked very much. I thought that the little glistening things on it were plastic, so I asked the salesman about them. Sensing the loss of a sale, he proudly explained: "No, sir, they are not plastic. They're acrylic." This, of course, prompted me to deliver a short lecture on plastics and why acrylics fall into this category. The salesman took it all very well, but a bystander blurted out that she would rather have a real chandelier than a plastic one. I bit my tongue and didn't tell her that this "fake," acrylic chandelier, with its beautiful angular pieces and modern design, cost roughly double what a "real," glass chandelier would cost.

High on Helium

Quickly now — what was the greatest dirigible tragedy in history? I'll bet just about everyone would assume it was that of the *Hindenburg*, which crashed and burned in a spectacular fashion when coming in for a landing at Lakehurst, New Jersey, in May of 1937. It wasn't. But it was the most spectacular. The giant dirigible had made ten routine roundtrips between Germany and the United States before the horrifying explosion. We still don't know exactly what happened, but the live description of the event provided by radio reporter Herb Morrison is mind-numbing: "It's crashing. It's crashing terrible. Oh, my, get out of the way, please. It's bursting into flames. And it's falling on the mooring mast. All the folks agree this is terrible, one of the worst catastrophes in the world. Oh, the flames, four or five hundred feet in the sky, it's a terrific crash, ladies and gentlemen. The smoke and the flames now, and the frame is crashing to the ground, not quite to the mooring mast. Oh, the humanity and all the passengers."

Morrison's report has a couple of fascinating features: it was the first recorded news report that NBC broadcast nationally; and, perhaps more interestingly, it does not appear to describe a hydrogen explosion. The *Hindenburg* was a dirigible, not a blimp. This means that it wasn't just a giant bag of gas, like the airships we see hovering over football games. Rather, it had a rigid aluminum framework over which a cotton skin was stretched. Inside were separate bags of hydrogen gas, which held the ship aloft. And what a ship it was.

The *Hindenburg* was the largest flying machine ever built — 804 feet long. It would have dwarfed a Jumbo 747 and was roughly the size of the *Titanic*. The crash Morrison described killed thirty-five of the ninety-seven people onboard, along with one crew member on the ground. Morrison's description, and the existing newsreel footage, portray a rapidly spreading fire, not an explosion. Witnesses spoke of flames that resembled a spectacular fireworks display; this is uncharacteristic of hydrogen, which burns with a virtually colorless flame. Some researchers have therefore concluded that the cause of the accident was not the igniting of the hydrogen but of the flammable cotton cover. It was common practice in those days to strengthen the cotton with iron oxide, cellulose acetate, and aluminum powder, a highly combustible mixture.

The theory these researchers put forward is that an electrostatic charge built up on the stretched cotton during a storm, and when the vessel's mooring lines were dropped there was a discharge through the metal frame, igniting the fabric. They tested surviving samples of the *Hindenburg*'s skin and found them to be extremely flammable. Indeed, even at the time of the disaster the Zeppelin Company, builders of the *Hindenburg*, may have thought that hydrogen was not the cause. They immediately took measures to reduce the flammability of the fabric that they were preparing for the *Graf Zeppelin*, the *Hindenburg*'s sister ship. They added a fireproofing agent, calcium sulfamate, to the skin and replaced the aluminum with bronze, which is far less combustible. They also reduced the electric potential between the skin and the internal structure by impregnating the ropes holding the fabric in place with graphite, a conductive material. The *Graf Zeppelin*, filled with hydrogen, went on to fly safely for millions of miles.

Publicly, the Zeppelin Company blamed hydrogen for the explosion, but they may have had a political card to play. At that point the United States was the only country that possessed supplies of nonflammable helium gas, but the Americans were unwilling to sell it to the Germans, fearing that they'd use it to construct airships for military purposes. Their worry was not unjustified, because the Germans had constructed over a hundred airships during World War I and dispatched them to bombard London. America had plenty of helium and had in fact constructed helium-filled dirigibles years before the *Hindenburg* accident. Four years before the *Hindenburg* exploded, the USS *Akron*, a military airship designed to carry small aircraft, crashed in a storm off the coast of New Jersey, killing sixty-nine of the seventy-two people aboard. That's twice as many as perished in the *Hindenburg* explosion. Tragically, a blimp sent out to look for survivors also crashed, and its entire crew died.

Accidents like those that befell the *Akron* and the *Hindenburg*, along with the advent of the airplane, spelled the demise of the great airships. But the Zeppelin Company, which, since the 1930s, has been pursuing other interests, is going back into the airship business. This time with helium. The *Zeppelin* NT, as the new semirigid dirigible will be called, is almost as long as a football field and will have swiveling propellers, one on each side and two at the rear. It will be almost as maneuverable as a helicopter and will be able to attain a top speed of ninety miles per hour. Only twelve passengers will fit into the cabin, but they will have a spectacular view. Zeppelin's idea is to use the airship for tourist excursions along the Rhine Valley. It won't be a cheap thrill, since the *Zeppelin* NT has a seven-million-dollar price tag.

But even those who can't afford a trip on the new Zeppelin may benefit from airship technology. The Japanese are designing blimps, which will float at an altitude of 65,000 feet and will be used to reflect waves used in cellular phone communication. The blimp's skin will be made of PVC coated with polyethylene, and the ship will be kept aloft for years. So we may yet see a revival of the great airships — with various new safety features. But, contrary to popular opinion, these great ships of the sky were relatively safe. Zeppelins flew for forty years before the *Hindenburg* accident, and they carried close to half a million passengers. There were no civilian casualties before the demise of the *Hindenburg*, and the chances of dying on a Zeppelin were about half that of being killed on a modern airplane.

Any sort of rebirth of the great airships is, however, contingent on the availability of helium. And therein lies a problem: we do not have an infinite supply. Helium is one of the products formed as uranium undergoes natural radioactive decay deep within the earth. Being extremely light, the gas diffuses upwards and mixes with natural gas, from which it can be isolated. In 1958 the U.S. Congress realized that a great deal of precious helium was escaping from natural gas wells into the atmosphere, and it allocated the then-astounding sum of one billion dollars to separate and stockpile helium.

The helium supply seemed secure, but then scientists discovered a whole new application for the gas. Helium boils at the lowest temperature of any substance (–269°C) and it never solidifies. Liquid helium is therefore the ideal substance to cool electrical wires and reduce their resistance to the flow of the current. It has allowed us to manufacture superconducting magnets, which have diverse applications, but probably none more useful than in magnetic resonance imaging. This technology affords doctors a noninvasive look inside the human body, and it is probably the most important medical diagnostic tool ever developed. The future of MRI, however, hinges on the availability of helium. Think about that the next time you fill those birthday balloons with the gas.

Harnessing Hydrogen

Queen Victoria watched attentively as Dr. Pepper picked up a seemingly empty bottle and announced, "And now the oxygen and the hydrogen will have the honor of combining before Your Majesty!" With that, he pulled the stopper from the bottle and pointed its neck towards an open flame. The queen and her entourage were astounded when a loud bang and a flash burst from the bottle. The hydrogen had indeed combined with the oxygen.

Dr. John Henry Pepper (not of soft-drink fame) was a chemist and the director of the Royal Polytechnic Institution in London, yet he was probably best known for his public lectures on science. He had been invited to entertain the queen with his optical illusions and "magic lantern" presentations on several occasions, but he'd lately been turning towards something slightly more dramatic. During the royal performance I just described he'd filled the bottle with a mixture of hydrogen and oxygen and explained to the monarch that the two elements could be made to combine; they would form water and release a great deal of energy in the process. Hydrogen could turn out to be a great fuel, Pepper went on, if only it could be obtained more easily. Alas, this was not possible.

The good doctor had made hydrogen by the method that had been first recorded by Robert Boyle, in 1671, in his *New Experiments Touching the Relation Betwixt Flame and Air.* Boyle, widely regarded as one of the fathers of modern chemistry, had explained how the addition of acids to iron filings "belched up copious and stinking fumes, which would readily take fire and burn with more strength than one would easily suspect." This reaction eventually captured the attention of young Henry Cavendish, one of the most bizarre and brilliant characters in the history of science. It was Cavendish who, in 1766, finally isolated this flammable gas and identified it as hydrogen. The name derives from the Greek term for water generator, since water is produced when hydrogen burns in the presence of oxygen.

Cavendish was a strange man. He looked like a mad scientist. His clothes — crumpled violet suit, frilled cuffs, and three-cornered hat — were shabby and seemed like relics from a bygone era. His voice was shrill, and he never looked anyone in the eye. Young Cavendish spent four years at Cambridge University but never earned his degree, because he simply could not face exams. For the rest of his life he would have trouble communicating with people, but in the laboratory none of that mattered, and his skills were unparalleled. Since he'd inherited a massive fortune, Cavendish never had to worry about working for a living. In fact, at the time of his death, he was the largest depositor in the Bank of England. So the scientist could have lived like a king, but instead he chose the life of a recluse and funded his scientific work with his inheritance. Cavendish's social life was virtually nonexistent. Although he had problems dealing with men, he would occasionally attend the scientific functions of the Royal Society. Women, however, were a different story. He instructed his maids to communicate with him only through written notes and to stay clear of him. After he accidentally met a maid on the stairs, he had a back staircase built for his exclusive use.

So, unencumbered by social obligations, Cavendish could work long hours in his home laboratory. It was here that he generated hydrogen by reacting iron or zinc with acids. He observed that when hydrogen burned in a closed container, water was produced. Water, therefore, contrary to ancient Greek dogma, was not an element: it could be made in the

laboratory. This was the final nail in the coffin of Aristotle's theory that everything was composed of air, earth, fire, and water. Cavendish had cleared a path for the progress of chemistry.

He had numerous interests and even experimented with passing an electrical current through different materials. Since no instrumental means to measure currents were yet available, he would estimate the current by grasping the ends of the electrodes with his hands and noting how far up his arm a shock would travel. The spectacle of an elderly man dressed in an antiquated suit clutching a pair of wires with his hair standing on end is surely one that audiences, including Queen Victoria and her entourage, would have enjoyed. But Cavendish steadfastly refused to exhibit his scientific skills in public. It was left to the likes of Dr. Pepper to introduce the world to the power of hydrogen.

Anyone who has ever seen a hydrogen explosion, even on a small scale, could easily recognize hydrogen as a potential fuel. In fact, way back in 1903 Constantin Tsiolkovsky suggested that liquid hydrogen would be an ideal fuel for launching rockets. This turned out to be prophetic; two stages of the giant

Saturn V, which took men to the moon, burned liquid hydrogen in the presence of liquid oxygen to achieve the tremendous thrust required. The space shuttle makes use of the same chemistry. Its huge external tank is filled with liquid hydrogen and liquid oxygen. Unfortunately the immense explosive potential of liquid hydrogen was tragically demonstrated by the *Challenger* disaster.

Liquid hydrogen not only lifts the shuttle into orbit, but it also produces the electricity needed during a mission. Have you ever wondered where the power inside the shuttle comes from? There are no onboard generators — not in the traditional sense, anyway. Power is generated by fuel cells. A fuel cell is a device that allows hydrogen and oxygen to combine without a combustion process. That's because the two reagents are separated by a solution known as an electrolyte and never come in contact with one another. Reaction occurs when electrons released from the hydrogen travel to the oxygen through an external circuit, leaving behind charged species (ions), which, with the aid of the electrolyte, combine to form water. The release and uptake of electrons actually takes place on the surface of a catalyst, usually platinum, which surrounds the hydrogen and oxygen electrodes.

As electrons travel through the circuit, they generate a current that can run any electrical device. The basic idea is that the energy derived from the combination of hydrogen with oxygen is in the form of electricity rather than heat. Furthermore, unlike batteries, fuel cells do not get used up. As long as hydrogen and oxygen are available, current is generated. The bonus, as far as the space program is concerned, is that the only product of the reaction is water. This is the water that the astronauts drink. No need to waste energy carrying bottles from earth.

By now, I'm sure, a question has arisen in your mind. If fuel cells produce energy without generating any toxic by-products, then why don't we use them here on earth? It's a good question, and the answer is that there are several problems involved. The catalysts for fuel cells are expensive; thirty thousand dollars worth of platinum was used in each fuel cell during the early days of the space program. By improving the technology, we have reduced this expense significantly, but the main problem — that of hydrogen availability — persists. Hydrogen does not exist in its elemental form, and so we have to manufacture it by heating methane or other hydrocarbons with water.

Right now the product of this process cannot compete in economic terms with petroleum, but as pollution control becomes more and more important to us, hydrogen will become more viable as a fuel. Vancouver and Chicago each have three hydrogen-fuel-cell-powered buses in full service, and the major car manufacturers are gearing up to put fuel-cell vehicles on the road by 2004. Then we'll have the honor of watching hydrogen and oxygen combine right before us.

Sticky Chemistry

At the Marine Science Laboratory in North Wales scientists spend a lot of time watching flies walk up a wall. No, they're not bored. Quite the opposite — they're very interested in finding out how these flies defy gravity and stick to the wall, because any insight into zygology can have very important practical applications. And what is zygology? It is the science of joining things together. Rivets, nails, screws, welds, and threads play an important role in our lives, but where would we be without adhesives? Planes would fall from the sky, furniture would disintegrate, tiles would drop from the walls, books and shoes would fall apart, and our teeth would go uncapped. Just imagine life without wallpaper, stamps, Scotch tape, or Post-It notes.

Any investigation of stickiness really has to begin with an examination of what goes on at the microscopic level, so the Marine Science Laboratory researchers first looked at the feet of flies under great magnification. Each of the fly's six legs is tipped with two little claws that grip irregular surfaces, but flies can also walk up and down smooth surfaces, like glass, with ease. Was there something else involved, the researchers wondered? Could a secretion of some sort play a role? They decided to see if flies left a residue when they walked on glass. They did. When the researchers treated a glass slide over which flies had walked with the stain Sudan Black B, which detects fatty materials, tiny footsteps showed up. Could this fat act as a glue?

Concocting an ingenious experiment to test this hypothesis, the researchers attached a thin tether to the back of a fly with superglue. Then they used a gauge to measure the force needed to lift the fly off the glass. Next they made the fly walk a plank covered with filter paper soaked in hexane, a solvent that dissolves fatty materials. Once more they placed the fly on a glass slide and hoisted the insect into the air. This time the force they needed was one tenth of that required before the hexane bath. Flies really do glue themselves to the wall. But what chemical explanation do we have for how this works? For that matter, how do we explain why gum sticks to hair but not to Teflon? And why is a sugar solution sticky while vinegar isn't?

Stickiness is a very complex business. Several factors are involved. The basic principle is that any two surfaces brought close enough together will adhere because of the attraction that exists between any pair of atoms or molecules. Atoms are made up of positively charged nuclei, where the mass is centered, and the negative electrons that orbit these nuclei. In a molecule, atoms can join to form chemical bonds, which are really made up of the electrons being attracted to both nuclei. But electrons are not stationary, and at any given moment a molecule or an atom will have a negative region where the electrons are found and a positive region where the nuclei are located. This momentary separation of charge is called a dipole. The negative end of one dipole can be attracted to the opposite pole of an adjacent one. This is called a Van der Waals attraction. When two surfaces are brought so close together that Van der Waals forces can be exerted between the surface molecules, we have stickiness. Each attractive force is small, but there are billions and billions of dipoles on each surface. If that is so, why can't we repair broken china by simply fitting the pieces together?

In theory we can, but in practice it is virtually impossible for us to bring two solid surfaces so close that they will exert Van der Waals forces on each other. No matter how smooth a surface may seem, it has peaks and valleys. A peak of just four hundred Angstroms (an Angstrom is 10^{-8} meters) means that surface-to-surface attraction becomes minimal, because Van der Waals forces are only exerted through a few Angstroms. Now, if one of the surfaces is mobile, its molecules can flow into the valleys and cover the peaks of the other surface, bringing the molecules close enough to feel each other's dipoles. Just think of what happens if you put a little honey between two fingers. The honey is mobile enough to flow into the crevices on both surfaces and get close enough to the surface molecules to trigger Van der Waals forces. But its mobility is not sufficient to produce a bond. Water also flows and forms Van der Waals bonds, but it isn't sticky. Another criterion must exist. And it does. To act as an adhesive, a material must not only stick to both surfaces, but its component molecules must also form strong links to each other so that they won't be separated when the two surfaces are pulled apart. Water may stick to both surfaces, but its internal intermolecular forces are weak, and stress easily pulls the molecules apart. Unless, of course, the water is frozen; two wet pieces of wood can certainly be secured together by freezing.

Honey, unlike water, has sugar molecules, which are strongly attracted to each other, as well as to other molecules. It is therefore sticky, but not sticky enough to use as an adhesive. For that we need a material that can easily flow to coat both surfaces, form good Van der Waals bonds with those surfaces, and then solidify to form a tough matrix of its own molecules, preventing their separation — just like our freezing water example. Naturally, we would like this to happen at ordinary temperatures, and this does happen with a simple glue made of flour and water. When wet, the paste is mobile, but as it dries the long starch molecules intertwine and become very difficult to separate. Protein molecules can also do the job. Remember how we used to send old horses off to the glue factory? The gelatin extracted from their hooves and hide was turned into an adhesive. Today, instead of starch or gelatin, we often use a synthetic material, polyvinyl alcohol (PVA) to remedy simple sticking problems. We can dissolve PVA in water to make white glue, a favorite of children. It flows well and hardens when the water evaporates.

An even better way to stick things together is to use an adhesive that can be applied in the form of small, very mobile molecules, which, through a chemical reaction, link together to form a matrix of giant molecules, or polymers. This is the way epoxy glues and superglues work. Epoxies consist of a two-component system, which reacts to form polymers, while superglues, like Krazy Glue, are applied in the form of small molecules called cyanoacrylates. On exposure to moisture in the air, these join together to make long, intertwined polymeric chains, which bind the surfaces together solidly. The effectiveness of this type of glue actually depends on the amount of moisture in the air. So, if someone asks you why Krazy Glue works better in Miami than in Phoenix, you can now tell them.

Obviously, one requirement for an adhesive is that it flow easily to cover a surface. This is a more complex business than it first appears. One might naively think that the governing feature is whether we are dealing with a thin or a thick liquid, but this is not the case. If we put a drop of oil in an iron skillet, it spreads, but on a Teflon surface it beads up. The explanation revolves around surface energies, which are a measure of the relative strengths with which atoms on the surface of a material are attracted to atoms inside the bulk of the material. In a sense, this determines how much attraction these surface atoms can spare for other substances. In the case of Teflon, very little. Teflon is composed of long chains of carbon atoms, with each carbon also joined to two fluorine atoms. The fluorines, which stick out from the carbon skeleton, represent the exposed part of the molecules, the part that could potentially interact with other molecules. Fluorine, once it has bonded to carbon, is notoriously unreactive, and it is not interested in forging other alliances. So, eggs — and everything else — will not stick to Teflon pans.

But we still don't have the whole story. Calculations show that even when a glue wets a surface effectively, Van der Waals forces cannot account for all of the strength with which the adhesive holds the materials together. There is another effect. As adhesives coat surfaces and harden, they trap tiny air bubbles in the microscopic crevices. These bubbles create a suction effect that has to be overcome when the surfaces are pulled apart. There are numerous glues on the market because surfaces can vary tremendously in chemical composition, and glues can vary dramatically in their strength as they harden. No single adhesive works on everything. And if one such universal substance did exist, how would we get the cap off?

We are always ready to learn more about adhesives. For example, we have started using cyanoacrylates instead of sutures on cuts. We also glue together fissures in the esophagus and use glue to treat cracked fingertips associated with eczema. Brain surgeons have used adhesives to reinforce weak points in blood vessels during surgery. But, like any other chemical, glues can create problems. Just ask the man who confused his cyanoacrylate bottle with his nasal decongestant. Luckily, the mucus in his nose kept the glue from sticking to tissue, and his doctors ended up removing a cast of the inside of his nostrils. Then there was the mother who mistook Krazy Glue for an antibiotic ointment and put it into her baby's eye. The eye had to be cut open under general anesthesia. And how about the Irish pub patron who sat on a toilet seat that vandals had coated with Krazy Glue? (One of those tabloid stories that's too good to check.) The man had a long time to contemplate the wonder of flies walking up and down the wall before he was discovered and transported to hospital — with the seat still attached. Doctors used acetone to dissolve the polycyanoacrylate, freeing the man from a very sticky situation.

Acetone is not the only solvent that can dissolve cyanoacrylate glues. Acetonitrile will also work, particularly for removing sculptured fingernails. While it is safe if used properly, ingestion can be lethal: cyanide is released when acetonitrile is metabolized in the body. In one terrible incident, a sixteen-month-old child accidentally swallowed fifteen to thirty milliliters of the solvent. His mother immediately called a poison-control center, but there was a misunderstanding about which chemical was involved. Somehow, the poison-control people received the impression that the child had ingested nail polish remover, or acetone. This is a less dangerous chemical, and emergency measures were not taken. The child was found dead in the morning, a victim of acetonitrile poisoning.

Glues themselves have been implicated in health problems. Remember the final episode of the 1995 season of *Seinfeld*, one of the most popular TV shows of all time? Jerry's friend George is about to get married. He and his fiancée, Susan, send out wedding invitations, but the wedding never comes off (much to George's relief), because poor Susan is poisoned. And it's all George's fault. The prospective groom purchases the cheapest envelopes for the invitations, and Susan is done in when she licks their toxic glue strips. The implication is that if George wasn't such an incorrigible tightwad, then this tragedy could have been avoided; more expensive envelopes would have been less toxic. Could this episode have been based on some real-life event? Hardly.

We subject envelope and stamp adhesives to stringent safety requirements. Since we're likely to swallow traces of the stuff, we have to regulate it as a food. Gum arabic from the acacia tree, dextrin from corn starch, and the water soluble resin polyvinyl alcohol are the adhesives we use most often. We also blend in other substances to enhance flexibility and spreading quality; these include glycerin, corn syrup, various glycols, urea, sodium silicate, and emulsified waxes. Preservatives such as sodium benzoate, quaternary ammonium compounds, and phenols are also included. These substances may not taste great, but they aren't poisons. In fact, cockroaches can survive for a long time on a diet of postage-stamp glue.

If we have any reason for concern about glues, then it has to do with allergic reactions. Some people may have sensitivities to the adhesives used in carpet backing, particle-board furniture, or even artificial fingernails. The latter presents an interesting case, one that I'm quite familiar with. A friend of mine developed typical allergic symptoms — including watery eyes, runny nose, and impaired breathing. Her allergist thought it was a reaction to pollen, but when her symptoms got worse he submitted her to a battery of tests. She showed no reaction to any of the typical allergens. One day my frustrated friend was bemoaning her lot to an old acquaintance, who thought to ask if she was wearing false fingernails. The acquaintance had experienced the same symptoms, but they'd disappeared when she removed her glued-on nails. My friend rushed home, took off the nails, and — you guessed it — by the next morning all of her symptoms had vanished.

I got involved at this point and suggested that we needed a controlled experiment to confirm my friend's suspicions. Being a pharmacist, she was game for a proper scientific investigation. When she put the nails back on, the symptoms came back. When she removed them, the allergic reactions disappeared. The allergist tested my friend for reaction to cyanoacrylate, the glue often used to paste on artificial fingernails. She reacted. Since then that allergist asks all of his patients (be they male or female) who show up with mysterious allergies whether they're wearing artificial nails. After I discussed this bizarre sequence of events on the radio, I had a call from another lady with a remarkable story. She had developed an allergy to acrylates while being fitted for an artificial tooth, and she later experienced an allergic reaction when walking by a nail salon. I'm not sure about the validity of this observation, but in the strange world of allergies anything is possible.

There is one more gluey story that merits telling. It's all about Spencer F. Silver. Granted, his name may not be a household word, but one of his inventions certainly is: Post-It notes. In 1998 the American Chemical Society presented its award for creative innovation to Silver, 3M's chief scientist, for inventing a glue that sticks when you want it to and releases when you want it to. And what a curious history this substance has.

Way back in 1968 Dr. Silver was working for 3M on pressure-sensitive adhesives. These are glues that bond instantly to a surface but can be removed without destroying that surface. Today we are very familiar with such products; peel-off stickers are everywhere. In 1968, however, they were virtually unknown. Scientists did realize that certain polymers, like natural rubber, could be peeled off under the right conditions, but they were not ideal. So Silver went to work. He investigated various synthetic polymers and eventually came up with one that was a weak adhesive and could be pulled off a surface. The difficulty was that it would not always pull off cleanly, and Silver lost interest.

Luckily, interest was rekindled by Arthur Fry, a chemical engineer working for 3M in the early 1970s. Luckily, Fry sang in a church choir. Luckily, Fry had tried to mark pages in his hymn book with pieces of paper that kept falling out. Luckily, Fry remembered Silver's weak glue. Luckily, he got his hands on some and used it to mark the pages in his hymn book with slips of paper that did not fall out and could be easily removed. These were the prototypes for Post-Its. It took Fry about a year and a half to work out the bugs. He developed a primer that could glue the adhesive to the paper, ensuring that it would not transfer to the surface to which it was applied. Then he invented a machine to make those familiar little pads. But they were not instantly successful — it took a clever marketing trick to launch them. Post-It pads were given to office workers in Boise, Idaho, and ninety percent of those workers ordered more when the freebies ran out.

Today's Post-Its are made with a very sophisticated technology. We can actually reuse them, because their adhesive is contained within thousands of little bubbles of urea formaldehyde resin. These bubbles break under pressure, but they don't all break at the same time. So, how many times can a Post-It be reused? I don't know. I should do an experiment. I'll leave myself a note to do just that. On a Post-It, of course.

LEARNING FROM THE PAST

Playing with Fire

Let's play a little word-association game. Chemistry. What does it bring to mind? Beakers? Formulas? Molecules? Maybe. But chances are that "Bunsen burner" rolled off many a tongue. There is probably no other piece of equipment as closely associated with chemistry as that ubiquitous little burner. Turn on the gas, adjust the air intake, and we're ready to simmer, stew, or boil. But what do we know of the man behind the burner?

Robert Bunsen was a professor of chemistry at Heidelberg University in Germany during the second half of the nineteenth century. He became interested in the study of arsenic compounds, and this interest would cost him dearly. Since arsenic derivatives are smelly and poisonous, Bunsen devised a face mask to protect himself from the nasty vapors. The mask had a glass shield and a long breathing tube which snaked out the window for fresh-air. Toxicity, however, wasn't the only problem Bunsen had to deal with; many arsenic compounds ignite and explode spontaneously in dry air. The chemist unfortunately hadn't taken adequate precautions in this regard. One of his samples exploded, shattering his mask and blinding him in one eye. Yet Bunsen, undaunted, pursued his chemical investigations.

Bunsen was a stickler for detail. One day he decided to analyze a sample of an ore to determine its beryllium content. At one point he set about filtering the final precipitate, which he had produced after putting the ore through a series of chemical reactions. The weight of this precipitate would be the key to the analysis. Much to his horror Bunsen saw a fly land on the filter paper and take off with some of the precious powder clinging to its landing gear. Bunsen's traumatized scream brought his students running. They quickly captured the fly and presented its corpse to the master, who cremated it in a platinum crucible. From the remains, Bunsen isolated the beryllium oxide with which the fly had absconded. After weighing the recaptured booty, he prepared to record a correct analysis for the beryllium content of the original sample. Then, since beryllium, like arsenic, is highly poisonous, it occurred to Bunsen to switch the focus of his work. He began to play with fire.

Why was Bunsen so interested in fire? Because laboratory workers had long been plagued by sooty, hard-to-control flames. As oxygen was necessary for combustion and soot was the product of incomplete combustion, Bunsen concluded that the secret to a clean flame lay in mixing the combustible gas with air in just the right proportion.

The prototype Bunsen burner consisted of a metal tube with strategically drilled holes through which air could enter and mix with the combustible gas flowing through the tube. A sliding metal cover allowed the operator to vary the number of open holes and thus control the character of the flame. Bunsen, however, never patented his invention. He did not believe that scientists should profit financially from their work. Research was its own reward.

Bunsen needed a clean flame because he had a passion for studying the brilliant colors he could produce by sprinkling various substances into a fire. He had noted that throwing sodium chloride (ordinary salt) into a flame always resulted in a bright orange-yellow glow. The same color appeared if sodium bromide, or any compound of sodium, was cast into the flame. Other elements also produced characteristic colors. In fact, Bunsen discovered the existence of the elements rubidium and cesium through the colors they produced.

Over a hundred years earlier Newton had shown how we can use a prism to separate white light into the colors of the rainbow. Bunsen now applied this principle to separate the colors of a flame into their individual components. The spectroscope, an instrument he developed with the physicist Kirchoff, allowed him to identify unknown substances by the colors they produced when heated in the flame of a Bunsen burner. Who cares what colors are produced in a flame? Just think of the glorious colors of fireworks. Or the bright red strontium flame of an emergency roadside flare. Or the yellow glow of a sodium vapor highway light. The original studies that led to these applications were painstakingly carried out by Robert Bunsen.

After toiling for a long time with spectroscopes and flames in the laboratory, the great man spent years writing up the results of his investigation for publication. The day he finished the manuscript, he left it on his desk and went out to celebrate. When he returned, Bunsen was horrified to see a smoldering pile of ashes where his treasured treatise had been. A flask of water sitting next to the manuscript had acted as a magnifying glass, focusing the sun's rays and igniting the paper. A lesser man would have surrendered to fate at this point, but Bunsen, despite his advanced age, rewrote the work. He eventually published the results of his spectroscopic research. The determined chemist certainly deserves to have his name mentioned in association with the word "chemistry."

A Man and His Elements

Let me own up to a little childish mischief. Way back in my student days I was intrigued by a story I had read about a chemistry professor at Glasgow University. The restrooms in that institution's chemistry building were on the ground floor, and the professor's office was on the third. But there was a secluded little balcony beside his office and along it ran a rain gutter; when pressed for time, the prof would relieve himself in the rain gutter instead of making a trip to the ground floor. Some of his students discovered his little secret and decided that it was just too good to ignore. They capitalized on their discovery in a manner appropriate for a chemistry professor. They placed a few pieces of sodium in the rain gutter in order to produce a memorable effect, and they succeeded. The professor, obviously a good sport, reported that the students' antics had resulted in a booming success. That's because combining sodium with water creates hydrogen gas and a great deal of heat; the hydrogen often ignites with some spectacular results.

Not knowing any professors who used rain gutters in this ingenious fashion, some fellow students and I resolved to torment innocent passersby. Montreal winters were ideal for this little adventure because snow often covered the sidewalks. So out we went with our little vials of sodium stored under oil and proceeded casually to drop a few pieces as we walked along. The snow reacted with the metal and produced neat little explosive bursts. People behind us thought they had wandered into a minefield. Years later, as I told this story during one of my chemistry lectures, one student eyed a chart hanging on the wall and proclaimed, with a gleam in his eye, "You should have used cesium, sir." He was right. Cesium is a more active metal than sodium, and it reacts far more vigorously with water.

That chart, which hangs in just about every chemistry

classroom in the world and lists cesium below sodium, is The Periodic Table of the Elements. And behind it hangs a tale — the tale of the unconventional but brilliant Russian chemist who first formulated it, Dimitri Mendeleev. Mendeleev was born in Siberia in 1834, the youngest member of a very large family. His first exposure to science was through the stories told to him by his sister's husband, a dissident scientist who had been sent to Siberia. The youngster showed such interest and ability that his widowed mother hitchhiked fourteen thousand miles to Moscow, where she attempted to enroll young Dimitri in school. Initially, no school would take him, but she persisted until he was accepted at the Pedagogical Institute in St. Petersburg.

Mendeleev was such an outstanding student that he was sent abroad to study in Paris and Heidelberg, the German city that at the time was the hotbed of chemistry. On his return to Russia, the bearded Mendeleev, who looked more like a caveman than a scientist, became professor of chemistry at the university in St. Petersburg and soon gained recognition as a leading light in chemical education. Like any good teacher Mendeleev attempted to organize the knowledge he had to impart in a systematic fashion. But, truth be told, chemical knowledge in that era was pretty basic. Professors simply asked their students to memorize what happened when chemicals were combined. They were responsible for knowing, for example, that when you dropped a piece of sodium into water, it ignited. So did potassium and cesium. A piece of aluminum did not. But nobody really knew why.

Convinced that elements did not react in a random fashion, that somehow their properties and behavior were systematic, Mendeleev wrote the names of the known elements on cards. Then he listed their various properties and the relative weights of their atoms. These atomic weights were a hot topic. They had intrigued Mendeleev ever since he had engaged in a discussion with the foremost expert in that field, the Italian chemist Cannizzaro, whom Mendeleev had met in 1860 at the famous Chemical Congress at Karlsruhe, Germany. This meeting had been convened in an attempt to impose some order on the vast amount of knowledge that chemists were accumulating. By 1868 Mendeleev had played his cards. He noted that if he placed them in horizontal rows, in order of atomic weights, with chemically similar elements arranged vertically beneath each other, then a systematic pattern would emerge. By virtue of their atomic weights, the elements could be grouped into families with similar properties. And then Mendeleev committed his boldest stroke. He predicted from holes in his periodic table that there existed elements that hadn't yet been discovered. He suggested that ekasilicon, an element with properties like silicon, would be found; and, indeed, in 1886 germanium was discovered.

When Mendeleev published his findings in the classic textbook *Principles of Chemistry*, it brought him fame and fortune. It also brought him scandal. Adoring students flocked around him, and in one case he returned that adoration, falling in love with a seventeen-year-old student named Anna Popova. This presented a serious problem, however, because Russia's most famous scientist was a married man. The marriage was not a happy one. When Mrs. Mendeleev accused her husband of being a bigamist, insisting that he was married both to her and to science, Mendeleev replied that he would happily become a monogamist and bed science alone. Apparently, he changed his mind when he met Anna Popova. Much to the consternation of the young lady's parents, Mendeleev divorced his wife and pursued Anna relentlessly. Mendeleev was determined to marry his new love, but, according to the religious law that prevailed in Russia at the time, a divorced man could not remarry for seven years. Mendeleev would have none of this and found a priest who, for the price of ten thousand rubles, was willing to defy the church and marry the happy couple. Due to the chemist's stature, no action was taken against him; the priest, however, was defrocked. When this inequity was pointed out to the czar, he meekly replied, "I admit Mendeleev has two wives, but I have only one Mendeleev."

Through all of this domestic turmoil the chemist continued to work hard. He was the very embodiment of his philosophy, "Look for peace and calm in work, you will find it nowhere else. Pleasures flit by, they are only for yourself. Work leaves a mark of lasting joy for others." When Mendeleev died in 1907, his funeral procession was dominated by students carrying huge periodic tables. The second Mrs. Mendeleev outlived her husband and became famous in her own right. After the Russian Revolution, when food was scarce in St. Petersburg and the government issued ration cards, Mrs. Mendeleev was given an extra allotment in recognition of her husband's contributions to Russia. She was easy to recognize, because she was one of the few portly women in the newly renamed city of Leningrad.

Leningrad has become St. Petersburg again, and the city is home to many educational institutions. One, in which chemistry is a leading discipline, has a facade adorned with the largest periodic table in the world. Inside, students are busy learning how to put chemistry to good use — perhaps they are even making ingenious use of sodium. They probably know what they're doing, unlike the American junior-high student who stole a piece of sodium from the school lab and wrapped it in a paper towel, which he then placed in his pocket. As he was walking home with a friend, smoke and then flames erupted from his pocket. They happened to be passing by a pond, so the quick-thinking friend pushed him into the water. Not a good idea. The sodium reacted vigorously with the water, and the boy was injured. So don't even think about creating mischief with cesium. Leave that to the experts. And, just in case they're active, steer clear of rain gutters on chemistry buildings.

Have a Drink for Dr. Snow

I'm not much for pub crawling, but I wouldn't have missed hoisting a brew at London's John Snow public house in honor of the gentleman after whom the establishment is named. As I did so I gazed out the window at the rather curious monument that stands in the little square in front of the John Snow. The monument is not a stately statue; it's an iron replica of a pump that stood in that very spot during the mid-1800s — the pump that drew the water that Dr. John Snow tied to a devastating cholera epidemic. Snow's demonstration of the link between impure drinking water and disease ranks as one of the greatest-ever contributions to public health.

Cholera is a dreadful disease. It causes diarrhea so severe that a person suffering from it can lose as much as ten liters of water in a day. If left untreated, cholera can lead to rapid dehydration and death within a few days. The disease first appeared in Europe in 1831, an import from the Indian subcontinent, where it was endemic. Over fifty-thousand Britons died within a year, sparking widespread panic. Physicians didn't know what to do. They plied their patients with arsenic and strychnine, they gave them tobacco enemas, they wrapped them in flannel soaked in turpentine, they bled them with leeches, and they blistered them with nitric acid. All, of course, to no avail. No one knew what caused the disease, but the prevailing opinion seemed to be that cholera was somehow transmitted by the bad air, or "miasma," that emanated from the sick and from garbage. One London dentist actually announced that the solution was to fire cannons every hour to disperse the bad air.

Snow, who was then a young apprentice doctor, attended to many of the epidemic's victims. He didn't buy the idea of poisoned air. Cholera, he was convinced, was the result of a poison that acted directly on the intestines and was therefore most likely introduced through the mouth. Snow's suspicion turned to water when he learned that the city of Birmingham had been spared. What was the difference here? Nobody drank the river water — it was so foul that nobody could stomach it. People in Birmingham drank only well water. By 1849 Snow had published a pamphlet suggesting that cholera was spread by water, but it was largely ignored. Having been trained through apprenticeship rather than one of the Royal Colleges, Snow was not a member of the establishment, and most prominent physicians looked down on his views. They didn't want their theories about miasmas challenged, and, in any case, by 1849 the epidemic had waned.

In 1854 Snow got his chance to prove his theory. A terrible cholera epidemic broke out in London. It was centered in the Soho district, and more than five hundred people died in the space of ten days within an area of a few city blocks. Taking a map of the city, Snow laboriously plotted out the locations of houses in which someone had come down with cholera. His efforts revealed an amazing pattern. The red dots signifying cholera cases clustered thickly along a major thoroughfare then known as Broad Street. What, on this particular street, could be causing cholera? Snow soon uncovered the answer. The pump that supplied the neighborhood with water sat right where the red dots were concentrated. Then Snow discovered that seven men who lived outside Soho but who had worked in the area around the pump all died of cholera; he also learned of the death of a woman who had moved from Soho to Hampstead but had sent someone to bring her a little of the Soho water she had been accustomed to drinking. But perhaps the most convincing observation Snow made was that in a nearby brewery, where the workers never drank water, not a single worker had contracted cholera.

At this point Snow approached city officials, showed them his map, and suggested that the handle of the pump in Broad Street be removed as a public health measure. Legend has it that the officials responded immediately, stopping the epidemic in its tracks. The true story is that by the time the pump handle was removed, the number of new cholera cases had slowed to a trickle. Snow did not singlehandedly stop the cholera epidemic, but that doesn't diminish his accomplishment. He proved to the world that cholera could be transmitted by polluted drinking water, and he went on to demonstrate that sewage was the culprit. Authorities eventually traced the origins of the Broad Street epidemic to a woman who had washed her baby's diarrhea-soiled diapers in water that was dumped into a cesspool that leaked into the well supplying the pump. Snow later determined that cholera was far more prevalent in homes supplied with water by a company that drew its water from an area near London Bridge than one that drew it further up the river. By 1855 he had proven that the disease was caused by an infectious agent in the water. The book published that year, *On the Mode of Communication of Cholera*, led to the chlorination of drinking water, a measure that saved millions of lives.

Robert Koch, the great German bacteriologist, eventually isolated the infectious organism that causes cholera from the corpses of victims and identified it as a bacterium. But not everyone believed that a tiny creature, visible only through a microscope, could cause such devastation. Dr. Max von Pettenkofer, one of Koch's severest critics, demanded that Koch supply him with a dose of bacteria large enough to "poison a regiment." The foolhardy von Pettenkofer then drank the concoction and gloated publicly when nothing happened. Koch responded by pointing out that not everyone who ingests the bacteria will come down with cholera. Stomach acid can destroy the bacteria, and people who secrete low levels of acid are the ones more likely to be affected. It seems that von Pettenkofer had an acid stomach to go along with his acid tongue.

The effect of acid on *Vibrio cholerae* was clearly demonstrated in the 1970s, when prison volunteers ingested the bacteria with, and without, sodium bicarbonate, a base that neutralizes stomach acid. Those who swallowed the bacteria with the bicarbonate chaser proved more vulnerable to the disease. These prisoners were not condemned to death by cholera. In the modern era we are able to cure the disease with antibiotics and oral rehydrating solutions. Sadly, these treatments are not always available in developing countries. Diarrheal disease due to various microbes in untreated water kills more people each year than AIDS or cancer. Health officials working to reduce the scale of the problem focus on introducing water treatment, on making rehydration solutions available, and on developing vaccines. One of the most exciting approaches to eradicating cholera — which still affects about five million people annually — involves the genetic modification of the potato.

Scientists have found a way to insert into a potato a gene that generates the protein the cholera bacterium uses to bind its toxin to cells in the gut. A person who eats this potato will form antibodies to the protein, antibodies that would recognize and destroy the binding protein should cholera bacteria be encountered. Based on animal studies, we can say that one such cooked potato a week for a month, along with periodic boosters, should do the job. It would be especially appropriate if they served such potatoes in the John Snow pub. Actually, potatoes would be more appropriate than beer. That's because Dr. Snow was particularly interested in the quality of drinking water since water was all he drank. Snow was a complete teetotaler.

The Man Who Lost His Head Twice

I sit here thumbing through a holy book. No, not the one you're thinking of. This one was written in 1789 by Antoine Laurent Lavoisier, the father of modern chemistry, and it is one of the great treasures of the McGill University chemistry department. The pages of *Traité Élémentaire de Chimie* are now yellow and brittle, the language is archaic, the diagrams are hand drawn, but Lavoisier's tome is probably the most famous chemistry book ever written. And there is a story behind it.

Lavoisier was an arrogant man. He was often accused of not giving credit to others, a criticism he shrugged off with the casual explanation, "The one who chases the hare is not always the one who eats the stew." But he transformed the science of chemistry from an archaic and eclectic array of theories and reactions to a modern, systematic discipline. And he lost his head not once, but twice.

Young Lavoisier followed in his father's footsteps, and in 1763, at the age of twenty, he became a lawyer. But it wasn't the law that captured Antoine's imagination, it was a popular-science lecture. A friend dragged him to the Jardin du Roi in Paris to watch the antics of scientific lecturer par excellence, Francois Rouelle. Lavoisier was absolutely mystified by the smoke, the flames, and the smells that highlighted the presentation. Chemistry, he realized, was his calling. Within a year Lavoisier had produced a scholarly paper on plaster of Paris and had helped to make a geological map of France. His scientific progress was so rapid that in 1768 he was elected to the prestigious Royal Academy of Science. Dabbling in science, though, was not a lucrative occupation, so Lavoisier set about securing his financial future by buying a share in the Ferme General, a private agency that collected taxes for the king.

The Ferme General was not popular with the people. Neither was Lavoisier's suggestion that they build a wall around Paris to discourage smuggling. Some claimed that the tax agency was actually attempting to poison the inhabitants of the city by keeping out fresh air and sealing in deadly exhalations. While this was clearly nonsense, it's odd that it was a study of air and exhalations that eventually brought Lavoisier everlasting fame. But first a little matter of personal chemistry had to be dealt with. At the age of twenty-eight the fledgling chemist fell in love with, and married, fourteen-year-old Marie Paulze. The union was an extremely successful one. Marie was interested in her husband's work and kept meticulous records of his experiments. She was also a gifted artist, and she chronicled his experiments in a series of detailed drawings that are reproduced in textbooks to this day.

One of these drawings illustrates the equipment Lavoisier used in an experiment designed to show that substances burning in air gain weight. He concluded that the extra weight must come from a component of air that is absorbed during combustion. Lavoisier called it "oxygen." He derived the name from the Greek term for "acid former," because he mistakenly believed that all acids contained this substance. But he wasn't wrong in his conclusion that air is composed of one-fifth oxygen, the "eminently breathable part of air," and four-fifths an inert material that he named "azote," from the Greek for "no life." We now know this gas as nitrogen.

Once Lavoisier had established that oxygen is the breathable part of air, he began to ponder what happens inside the body when we inhale it. One of his most famous experiments involved a guinea pig that he kept under an atmosphere of oxygen. Lavoisier determined that in the process of respiration oxygen is converted into "fixed air," soon recognizing this as carbon dioxide. Furthermore, by surrounding the guinea pig with ice, he showed that the production of carbon dioxide was accompanied by the melting of ice. Heat was being released from the animal's body. When another experiment showed that by burning charcoal, a relatively pure form of carbon, one could produce the same proportions of carbon dioxide and heat as the guinea pig, Lavoisier concluded that the animal's body "burned" food to generate body heat. The experiment also expanded our sense of the term "guinea pig"; forever after it would refer to any being that is used as the subject of an experiment.

Lavoisier did, in fact, decide that he needed a human guinea pig, and one of his assistants fit the bill nicely. As illustrated in a famous sketch by Mme. Lavoisier, the assistant wore a mask that allowed him to inhale oxygen and then breath into an apparatus designed to collect and identify the components of the exhaled air. Humans, too, it appeared, inhaled oxygen and exhaled carbon dioxide. Life involved some sort of combustion process.

Over the next ten years Lavoisier unraveled many of the mysteries of this slow combustion process. He showed that if a body was performing physical labor, then oxygen intake increased, as it did when the body was subjected to cold temperatures. Food, it seemed, was being burned to supply the needed energy, and this required oxygen. Even when the body was completely at rest, it was taking in oxygen and releasing carbon dioxide. Combustion was still meeting the energy needs of a beating heart, expanding and contracting lungs, and myriad other bodily processes. This energy requirement came to be known as the "basal metabolic rate," one of the cornerstones of the modern science of nutrition. Lavoisier had fathered not only chemistry but also nutritional science.

The brilliant Frenchman's legacy will be with us forever. *Traité Élémentaire de Chimie* was the world's first real chemistry text. In it Lavoisier introduced a whole new system of nomenclature, which we still use. No longer would chemists refer to "oil of vitriol" or "flowers of zinc." Instead they adopted names like "sulfuric acid" and "zinc oxide," names that reflected the actual composition of the substances in question. Lavoisier clearly defined elements as substances that could not be broken down further by chemical means. Chemistry was evolving into an organized science!

Accolades and riches should have rained down upon Lavoisier in recognition of his accomplishments, but it was not to be. His early involvement with the Ferme General came back to haunt him. In 1793, during the Reign of Terror that came on the heels of the French Revolution, Lavoisier, along with other members of the hated Ferme, was arrested on a trumped-up charge of having "mixed water and other harmful ingredients into tobacco." On May 8, 1794, Lavoisier was tried, found guilty, and guillotined. A fellow scientist who observed the tragic event commented that it "required only a moment to sever his head, and probably one hundred years will not suffice to produce another like it."

A century after Lavoisier's death the city of Paris finally erected a statue in his honor. But the sculptor made a grave error: he modeled the head after a bust displayed in the Academy of Sciences that he believed was of Lavoisier; it wasn't so, for the second time, the great scientist lost his head. Unfortunately even the wrong-headed statue no longer exists. During World War II the invading German forces melted it down to make bullets. Thankfully, one monumental work of art still stands as a testimonial to the contributions of the great chemist. Jacques Louis David, perhaps best known for his painting *The Death of Marat*, has left us a wonderful portrait of Lavoisier. The painting depicts Antoine with some of his gas-collecting equipment, Marie by his side. It is among the most prized possessions of the Metropolitan Museum of Art in New York City. Every lover of science should make a pilgrimage there to admire this fantastically lifelike portrait and to give homage to the father of modern chemistry.

You've Got to Hand It to Ignaz

The custom of kissing on both cheeks as a form of greeting has always bugged me. Microbugged me. Do we really need to punctuate an encounter by exchanging oral microbes? What's wrong with a good, old-fashioned handshake? Well, plenty. Hand-to-hand contact may actually be an even more efficient way of transmitting bacteria and viruses than mouth-to-face, or even mouth-to-mouth, contact. The numerous, fat-laden crevices on the surfaces of our hands are an ideal breeding ground for microbes. Shaking hands and then touching your nose or your eyes is a great way to get infected. In fact, you're more likely to get bugged in this fashion than you are if someone kisses you or coughs in your face. This is not just wild speculation. Thanks to sixteen self-sacrificing people, we have the data to back it up. They generously volunteered to smooch for a minute with a cold-infected partner so that researchers could study the transmission of the cold virus. Only one volunteer caught a cold. But when mutual hand stroking was called for, the colds spread like wildfire — unless the volunteers frequently and scrupulously washed their hands. Ignaz

Semmelweiss would have been thrilled by the results of this study.

Who was Ignaz Semmelweiss? Only one of the greatest figures in the history of medicine, the man who discovered that the simple process of hand washing could save lives. Too bad he wasn't able to convince his contemporaries of the significance of his discovery. Semmelweiss, who was born in Hungary, graduated from the University of Vienna Medical School in 1844. He then became an assistant to the highly respected Dr. Johann Klein, a professor of obstetrics. Young Dr. Semmelweiss quickly discovered that the joy of childbirth was often followed by tragedy. Numerous women died within a week of giving birth from a disease known as childbed fever. Nobody knew what caused it, but many were of the opinion that the culprit was some sort of infectious, weather-related vapor in the air. This seemed to make sense to most physicians, because the death rate had a seasonal variation; but it didn't make sense to Semmelweiss. He had observed that the death rates varied quite a bit between the hospital's two geographically separated obstetrics wards. Surely, the air quality didn't change from one part of the hospital to the next. Furthermore, why was it that many more women died after giving birth in the hospital than after giving birth at home?

It was a strange business. In the "death ward" births were attended to by highly trained doctors, whereas midwives attended to the births that took place at home and in the ward where the incidence of childbed fever was much lower. Semmelweiss became obsessed by this conundrum. He performed autopsies on a number of the dead women, searching for some causative agent. He found none. When his breakthrough came, it was prompted by tragic circumstances. One of his colleagues cut himself during an autopsy and quickly died of symptoms that were remarkably similar to those of childbed fever. Semmelweiss surmised that some sort of cadaver particles must have entered his friend's bloodstream, killing him. And perhaps similar cadaver particles were also killing the women in childbirth. Now the difference between the two obstetrics wards became clear. The doctors who assisted in the "death ward" births and performed internal exams on the women before and after they delivered often came directly from the autopsy room where they were trying to solve the horrific problem of childbed fever. Could they be infecting their patients with some sort of cadaver particles? After all, their hands constantly smelled of cadavers. Semmelweiss could even rationalize childbed fever's seasonal variations this way. He determined that the rate rose when a new crop of students keen on doing autopsies entered the medical school and fell when those students were studying for exams.

The answer now appeared obvious. Semmelweiss urged all doctors and students to wash their hands thoroughly after performing autopsies. But even with thorough washing a faint smell of the autopsy room persisted, so Semmelweiss added the instruction to rinse the hands in a chlorine solution. Chlorine was already known to eliminate smells, although no one understood how it did so. The results of Semmelweiss's hand-washing campaign bordered on the miraculous. Within a year the obstetrics ward's death rate fell from a high of thirty percent to just three percent. The notorious "death ward" was no more. Semmelweiss was elated by this result, but he was also troubled by it. He realized that he, himself, had probably been responsible for many deaths as he rushed back and forth between the obstetrics ward and the autopsy room. These feelings of guilt, compounded by his conviction that he had made a major discovery, transformed Semmelweiss into a hand-washing zealot. This didn't sit well with many of his Viennese contemporaries, who resented being called murderers by the Hungarian outsider. Indeed, they made his life so miserable that he fled to Hungary to take up a post at a Budapest hospital. There he brought about the same kind of childbirth death-rate drop he'd accomplished in Vienna. And the Viennese death rate started to climb steadily in Semmelweiss's absence.

Historical accounts often suggest that Semmelweiss was a medical martyr driven out of Vienna by establishment physicians who refused to accept anything new. That wasn't exactly so. In a fashion, Semmelweiss was the architect of his own misfortune, because he never published his results or the details of his procedures. His critics were therefore unable to evaluate his data; all they had was his browbeating rhetoric. And those who tried to follow in his footsteps often failed, because they did not realize how scrupulous the hand washing had to be. Because of arrogance and poor communication, Semmelweiss's efforts, as significant as they were, never amounted to a true scientific breakthrough. Haunted by the rejection of his theories, the pioneer of antisepsis ended up in a mental asylum. He wasn't there for long, though: two weeks after being admitted, Semmelweiss was dead. Popular lore has it that he died of an infection he acquired during one of the last surgical procedures he performed, an infection just like the one that caused childbed fever. The historical facts, however, indicate that he likely died after asylum attendants beat him. In any case, this was not a fitting end for the man who saved many a life by merely washing his hands.

Vindication finally came in 1879, during a scientific meeting. A French obstetrician was subjecting the participants to an anti-Semmelweiss harangue when a short man stood up, went to the blackboard, and drew a picture of a streptococcus bacterium. "This," he said, "is the killer that Semmelweiss eliminated." The man was Louis Pasteur. Case closed. Soon doctors around the

world were washing and scrubbing to protect their patients and themselves.

Today hand washing is more important than ever. Many hospital-acquired infections are of the MRSA type (methicillin resistant staphylococcus aureus), which are difficult to deal with. Medical workers who practice good hygiene can reduce the risk. But everyone, of course, should pay close attention to hand washing. It may be one of the most effective disease prevention measures we can take, since microbes lurk everywhere. You would be amazed at what can be cultured from toilet flush levers, faucets, telephones, and other people's hands and cheeks. Maybe those affected types who gurgle "Dahling!" and air kiss when they meet are on to something.

Mesmerizing Medicine

Let us go back for a moment to the Vienna of the late 1700s. While Mozart was the toast of the town, the city was also abuzz with stories about the exploits of Franz Anton Mesmer. Dr. Mesmer, who had studied medicine at the University of Vienna, was now rumored to be curing people in the most unorthodox fashion.

Mesmer had become interested in the work of a certain Professor Maximilian Hell, who claimed to be able to cure people with magnets. This idea appealed to the young physician, who in 1766 had written a thesis on the effect of the stars and planets on human health. In the thesis he had invoked an old notion about the universe being filled with some sort of invisible fluid that connected people to the planets and to each other. Could it be, Mesmer now wondered, that this fluid was of a magnetic nature, and inadequate supplies of it in the body led to disease? Had Hell stumbled on a way of restoring the magnetic fluid by using external magnets? It was worth a try.

So Mesmer devised a system for testing restorative magnetization. Patients sat around large tubs of either iron filings or magnetized water and gripped metal rods that protruded from them. Tales of Mesmer's success spread rapidly through Vienna, and the afflicted rushed to consult Mesmer — much to the consternation of the medical establishment, which considered the treatment to be outright quackery. Not that the establishment physicians could provide any effective alternatives: their purging and bloodletting were, if anything, more dangerous than Mesmer's magnetic therapy. The establishment, however, prevailed, and Mesmer was forced out of Austria.

Paris proved to be a friendlier environment, and soon Parisian socialites were flocking to Mesmer's elegant healing salon. It must have been quite a sight to behold: the master in his long purple robe embroidered with astrological symbols floated around the room, exhorting the healing power of the magnetized rods. Then it occurred to Mesmer that the rods were not essential for his cures; in fact, his own hands worked just as well. His rationale? "Animal magnetism" could be transferred from body to body, replenishing deficient supplies.

Next, Mesmer decided that even he wasn't necessary. He could train others to effect those animal-magnetism cures. Since his clientele was mostly female, Mesmer hired handsome young men to sit opposite the ladies, knee to knee, and "magnetize" them by massaging their shoulders and breasts. In some cases the ladies fainted and had to be carried to private recovery rooms, where more "curative" work could be performed. As tales of the wonderful cures spread, Mesmer grew wealthier and wealthier.

Once again the scientific establishment began to take notice, and jealousies developed. In 1784 a royal commission, which included some of the most famous scientists of the time, examined Mesmer's supposed cures. Benjamin Franklin, then the American ambassador to France, was a commission member, as was the noted chemist Antoine Lavoisier. Dr. Joseph Guillotine, of head-chopping fame, joined them. This time, however, the heads only rolled figuratively; the commission found Mesmer's claims lacking in substance and asked him to leave Paris. The verdict was: "The imagination without the magnetism produces convulsions, and the magnetism without the imagination produces nothing."

In debunking Mesmer, the commission had touched on the crucial role of the imagination in governing health. Indeed, what Mesmer had observed was not animal magnetism at all, but rather the placebo effect. Today we know that forty to fifty percent of people suffering from an illness will respond to the power of suggestion, often in a remarkable fashion. Although Mesmer's reputation was badly tarnished by the findings of the Royal Commission, his exploits laid the foundations for the science of hypnosis, and Mesmer himself will forever be commemorated by the term "mesmerism."

Mesmer was certainly the most famous magnetic healer of his time, but he was not the only one. In England James Graham was busy peddling his unique brand of magnetic malarkey. Graham's performances were positively theatrical. He decorated the hall where he delivered his "healing" lectures with the crutches cast away by those he had "cured." The audience would wait for the performance to begin in great anticipation. Finally, to the sound of music, Dr. James Graham would appear through a trapdoor. Not your usual entrance for a lecturer, but this was no ordinary lecturer. The year was 1779; the place was the newly opened Temple of Health situated on the banks of the Thames in London. The people had come to hear about a novel technique to cure disease that involved magnets and electricity.

James Graham had studied medicine in Edinburgh and then spent two years in Philadelphia, where he became acquainted with Benjamin Franklin's experiments in electricity. He was intrigued with the possibility of applying this mysterious force to the field of medicine. His crafty mind began to formulate an idea. If doctors could claim to cure illness through bloodletting, cupping, or foul purgatives and emetics, why not through magnets or electricity? The plan for establishing a Temple of Health was hatched.

This impressive establishment housed not only the lecture room where Graham delivered his unsubstantiated lectures on electrical healing, but also a number of chambers equipped with the expensive electrical equipment, which supposedly delivered the goods. Patients could choose between a "magnetic throne" or an "electrical bath" to alleviate their problems. For those who were not into high tech there was the "earth bath" — essentially, a mud wallow. If none of this appealed, then clients were invited to purchase nostrums. A concoction with the subtle name Elixir of Life could be had for an exorbitant price, but the Temple of Health was not designed to accommodate the poor.

The central attraction at Graham's temple was the Celestial (or, as he sometimes called it, the Magnetico-Electro) Bed. For one hundred pounds a night, childless couples could rent the elaborately decorated bed and frolic amorously while being exposed to the effects of the seventeen hundred pounds of magnets built into the frame. Graham promised superior ecstasy and guaranteed conception.

For a little while the Temple was the place for the fashionable to see and be seen. Some came out of curiosity, some were prompted by testimonials from cured patients. And, undoubtedly, some of Graham's patients were cured. Some suffered from self-limiting diseases, and others were restored to health by the powerful placebo effect. But then people began to realize that Graham's success rate was no higher than that of any other purveyor of strange and wondrous therapies. The Temple's popularity waned, and Graham's ego crumbled. The "doctor" could not cope with his fall from the pedestal, and he went mad. He took to doping himself with ether vapor and running through the streets, sometimes stripping off his clothes to give to a beggar. James Graham died insane at the age of forty-nine.

Maybe he should have slept with a magnet on his head. At least that's what William Philpott, an American psychiatrist, might have recommended. Dr. Philpott recently carried out tests on over three thousand patients who agreed to sleep with magnets (equaling the strength of about four little fridge magnets) strapped to their heads. He reported that his subjects felt more energetic and less depressed; they also experienced an enhanced libido. Philpott hypothesized that the magnets stimulate the pineal gland to produce more melatonin, which he linked with the positive effects. This curious research was never subjected to peer review, but seems harmless enough. I'm always game to try new scientific ventures, so I gave magnets a shot. All I got was a headache. Maybe I should have my head examined.

The Chew-Chew and Do-Do Man

In the late 1890s a number of scientists around the United States received a strange package in the mail. The contents, labeled "economic ash," were actually samples of human excrement. The sender was Horace Fletcher, one of the most famous dietary gurus of the era. His motive? To show these scientists that proper digestion produced stools, "with no stench, no evidence of putrid bacterial decomposition, only with the odor of warm earth or a hot biscuit." And what was the secret to proper digestion? To eat like Horace Fletcher; or, as one of his famous devotees, Dr. John Harvey Kellogg, said, to "fletcherize."

Horace Fletcher was born in 1849, and he spent his youth traveling around the world. By the time he turned forty he had made himself a rich man by importing Japanese art, toys, and novelties. His portly appearance bespoke the good life. And then came a pivotal moment: Fletcher applied for life insurance, only to be turned down due to an unfavorable medical exam. He then became obsessed with health. The man who, up until that point, had shown no interest in matters of science, now began to expound a theory of health. "Troubles come from too much of many things," he said, "among them too much food and too much worry."

We could counter worry, Fletcher insisted, with the process of "menticulture." Today we would call it stress management. But Fletcher advised us to combat the nutritional causes of disease in a much more bizarre fashion. The secret to good health, according to Fletcher, was mastication. If we ate too much food without chewing it thoroughly, we would suffer indigestion and fail to assimilate the nutrients properly. But if we chewed our food until it liquefied in our mouths and then let it slide down the gullet, we would avoid the ravages of disease. So Fletcher chewed and chewed, once chewing 722 times to liquefy a shallot. He claimed that those who employed this method would be satisfied with far less food and require only twelve to fifteen mouthfuls to be satiated. Fletcher himself found that he could easily get by on about sixteen hundred calories, far less than the average daily consumption at the time. Particularly noteworthy was the fact that this represented a considerably smaller protein intake than doctors were recommending.

Although Fletcher was initially regarded as a crank, people started to pay attention when the nutritional apostle demonstrated amazing feats of strength in spite of his low calorie intake. He beat younger men in bicycle races and routinely bested the strongest university students in weight-lifting contests. So America began to "fletcherize," and scientists studied the results. Professor Russel Chittenden of Yale University concluded that fletcherizers took in far less protein than what was considered to be the optimal amount, and yet their strength improved. Chittenden also maintained that his own headaches and rheumatic knee pain were resolved by the chewing regimen.

Fletcher's popularity grew and grew, and more and more people started to chew and chew. The Great Masticator, as he was called, spoke to medical societies; his books were translated into many languages and his theories were discussed in the foremost medical journals — especially his theory that there would be no slums, no degeneracy, no criminals, and no need for doctors if only the world would chew properly. And for a while the world tried. Urged on by the likes of Thomas Edison, John D. Rockefeller, and Kellogg, people exercised their mandibles like never before. They stopped eating meat and switched to fruits and vegetables, because meat was hard to liquefy. They ate less and felt better. The slender look became chic. But gradually people tired of all that chewing, and the memory of the man who had captivated a whole generation began to fade.

Perhaps Fletcher bit off a bit more than he could chew, but his ideas were not completely nonsensical. Since saliva contains enzymes that begin the process of breaking down food components, proper chewing does aid digestion. Properly digested food is more readily absorbed, so, in theory, we could get away with eating less. People with inflammatory bowel disease may also limit digestive irritation by thoroughly chewing their food. And there is yet another benefit, especially to those afflicted with flatulence. Extensive mastication means less swallowed air and, consequently, fewer emissions.

Although modern science cannot fully endorse Fletcher's emphasis on mastication, we must admit that Fletcher was one of the first to rail against dietary excess and the first to demonstrate that our daily protein requirement is actually quite modest. As far as his theory that proper chewing prevents criminal behavior goes — well, it's a bit hard to swallow. But since we haven't been too successful in this area, perhaps we could try a little fletcherizing anyway. It wouldn't hurt. We might even lose some weight in the process — a study conducted at the prestigious Mayo Clinic in Minnesota supports that notion.

Researchers were inspired by the knowledge that cows expend a fair amount of energy by chewing. About twenty percent of the calories cows consume they burn through rumination. How do we ascertain this? By comparing the energy expenditure of cows fed intravenously to that of cows fed normally. In other words, cows that don't have to use energy to chew their cuds are more likely to put on weight. The researchers wanted to know whether there was a human parallel — that is, could humans possibly lose weight by chewing? They fitted seven adults with masks that allowed inhaled oxygen and exhaled carbon dioxide to be measured, since calorie expenditure can be calculated from the ratio of these two gases.

The volunteers were then asked to sit in a dark, silent, temperature-controlled lab with their arms and legs supported; they would not have to exert themselves doing anything that wasn't required by the experiment. The researchers measured the subjects' energy expenditure at rest for thirty minutes, and then they gave the subjects calorie-free gum. As reported in *The New England Journal of Medicine*, the subjects chewed the gum "at a frequency of 100 Hz, a value that approximates chewing frequency at our institution." A metronome kept everyone chewing at a constant rate. Just like the cows, the human subjects increased their energy expenditure by nineteen percent through chewing. What does this mean? The Mayo Clinic team concluded that by chewing gum during waking hours a person could drop five kilograms in one year. But, according to their data, we'd have to chew at a pretty rapid pace to achieve such a weight loss. One hundred Hz, in everyday language, means a hundred times a second. Now, either Mayo Clinic people are an especially athletic bunch of chewers, or the researchers forgot their elementary physics. I suspect the latter. Surely, they meant a chewing frequency of a hundred times a minute, not a second.

An interesting bit of research to be sure, but I would still recommend other forms of exercise for weight loss. If you aren't into exercise, though, you may want to take a look at a study carried out at Johns Hopkins University. Here researchers examined how various aspects of people's environments affected their eating habits. They discovered that when walls were colored light blue or green, forty percent of people chewed their food more and ate less. When the walls were red, yellow, or orange, chewing rates went down and food consumption went up. Maybe that's why so many fast food restaurant designers choose these colors. Horace Fletcher would not approve. He would probably chew them out.

The Catty Dr. Katterfelto and His Noxious Insects

These days everyone knows that bacteria and viruses can make us ill, but the idea that many ailments are caused by living microorganisms is relatively new. Up until the middle of the nineteenth century most people believed that influenza was caused by unfavorable planetary influences and that malaria was the result of bad air. Only in the late 1800s did Louis Pasteur begin to unravel the mysteries of microbes and their relationship to disease.

Although Pasteur usually gets credit for introducing the bacterial theory of disease, the first person to think along these lines was a popular conjurer called Dr. Katterfelto, one of the most unusual characters of the eighteenth century. Katterfelto was of Prussian origin, and he never quite mastered the English language. Nevertheless, in 1781 he introduced Londoners to a brand-new style of entertainment: a blend of science and magic. His show featured scientific demonstrations, conjuring tricks, and discussions of health.

The demonstrations involving electricity, hydraulics, and chemistry must have seemed truly magical to the uninitiated, and this suited Katterfelto just fine. He saw no need to destroy the illusion by providing explanations. In fact, he did nothing to dispel the myth that his everpresent black cat was actually responsible for the "magic," and the good doctor was not opposed to selling his pet's offspring "in order to propagate the breed of this wonderful cat." But the creature that brought Katterfelto enduring renown was not the black cat; it was a microorganism that could only be seen under a microscope. Anton van Leeuwenhoek, a Dutch lens grinder, had invented this instrument about a hundred years earlier, and with great excitement he described the little "animalcules" he had seen when he examined a drop of water through it. Leeuwenhoek had even noted that the little beasts were present in his saliva and had cleverly observed that they were killed by hot coffee. But, of course, he did not recognize the minute creatures as bacteria.

Conjurers are quick to seize scientific discoveries and incorporate them into their performances. So it was with Katterfelto. People lined up to peer through his solar microscope and catch a glimpse of the cavorting "insects," as he called them. But it was during the London influenza epidemic of 1782 that our hero demonstrated his cleverness most compellingly. Londoners read in their newspapers that at a Katterfelto performance, "the insects on the hedges will be seen larger than ever, and those insects which caused the late influenza will be seen as large as birds, and, in a drop of water the size of a pin's head, there will be seen above 50,000 insects; the same in beer, milk, vinegar, flour, blood, and cheese. There will be seen many surprising insects in different vegetables."

What the people saw were various bacteria and other microorganisms, not the viruses that cause influenza. Still, Katterfelto was on the right track when he linked microbes to disease. After being terrified by these tiny active creatures invisible to the naked eye, audience members were treated to Katterfelto's hocus-pocus, his black cat, and his scientific tricks. They were astonished, and they readily concluded that the strange man in the long black coat and velvet hat was blessed with remarkable powers.

When Katterfelto informed his audiences that he possessed a cure for influenza, they bought it. But the only real effect of "Dr. Bato's Remedy," as he called his nostrum, was that it riled the physicians of the day, who as yet had no means to combat the virus. They criticized and mocked Katterfelto vigorously, but the conjurer was not one to take it lying down. He planned an ingenious revenge against his detractors. Katterfelto declared that he'd received letters from apothecaries, physicians, and surgeons imploring him to set free the "noxious insects" in his possession in order to increase the number of cases of influenza. Why? Because these unscrupulous men "preferred their own pockets to the health of His Majesty's subjects." Naturally, Katterfelto promised never to commit such a dastardly deed; he claimed to have "reserved those noxious insects for the express purpose of exhibiting them by his solar microscope." The lines of people waiting to see their doctors shrank as the lines of people anxious to view the influenza-causing insects held captive by the heroic Katterfelto grew longer and longer.

And what became of this Prussian charlatan when an influenza epidemic broke out — an epidemic that Dr. Bato's Remedy could do nothing to curtail? For a time he was publicly ridiculed — and he was even lampooned as "Dr. Caterpillar" in a play appropriately titled *None Are So Blind as Those Who Won't See*. But he did not relinquish his own theatrical ambitions. Instead he waited for some new scientific phenomenon that he could incorporate into his act. Finally, the Montgolfier brothers introduced the hot air balloon, and Katterfelto had his new demo. During his show he launched what he called "fire balloons" and explained to the audience the workings of this new scientific marvel. At least until one of the balloons drifted off and ignited a haystack in a farmer's field.

Katterfelto did not have enough money to compensate the farmer for the fire damage, and he landed in jail. On his release he toured the countryside with a show that featured yet another novel effect. Katterfelto had learned about magnets, so he fitted his daughter with a metal helmet, which he secured to her body with straps that went under her armpits. Using a huge magnet, he raised the girl to the ceiling. But the magician was never able to recapture the fame he had once attained by revealing the dreaded "animalcules" to people through his solar microscope. He spent the rest of his life as an itinerant performer, constantly in debt. Dr. Katterfelto may not have been the most knowledgeable or savory character, but he did manage to link disease to microbes some eighty years before Pasteur conducted his momentous experiments.

Conquering with Conkers

There were no video games. There was no Internet. Television had not yet been invented. So in the early 1900s British children played conkers, a game that is perhaps best described as an early form of *Survivor*. The winner was the player whose conker survived the longest in battles with other conkers. So what was a conker? A horse chestnut, tied to a string, which a child could swing around and smash into an opponent's chestnut. Competitions were fierce, and winning nuts were highly prized. Apparently, though, some resourceful players didn't play cricket. They cheated — chemically. Some genius discovered that immersing a chestnut in acetic acid disrupted its cell structure, causing the nut to release its water content. The next step was to harden and preserve the parched conker with formaldehyde. But, just as this intriguing chemical practice was achieving wide popularity, the government stepped in. It wanted all available conkers for the war effort. No, they were not planning to barrage the Germans with chestnuts. Well — not directly, anyway.

The enlisting of horse chestnuts for duty started in 1917, when the Director of Propellant Supplies sent letters to British schools asking students to gather the nuts to help win the war. This was a puzzling request, to be sure. We need a bit of background to understand what prompted it. At the time, the main propellant used by the British army was a substance called cordite, which had replaced old-fashioned black powder. Unlike black powder, cordite burned without producing any smoke, allowing machine gunners to see their targets clearly and snipers to fire without giving away their positions. Ballistite, an early smokeless propellant, had been discovered way back in 1888 by Alfred Nobel, who had mixed gun cotton (nitrocellulose) with nitroglycerin. Cordite was an improved version, made by adding Vaseline and acetone to the mix. The Vaseline helped lubricate the gun barrels, and the acetone functioned as a solvent, permitting the components to be blended intimately and extruded through a die. As the mixture emerged from the holes in the die the acetone evaporated, leaving the manufacturer with long strings of cordite.

In Britain, the manufacture of cordite had commenced in 1889 in the royal gunpowder factory at Waltham Abbey. The acetone, which was critical to the process, was made from the distillate collected from wood that was heated to a high temperature. The best wood for this purpose came from the forests of continental Europe, and it was therefore unavailable to the British after the start of World War I. But in 1915 a chance meeting solved this problem. C.P. Scott of *The Manchester Guardian* introduced David Lloyd George, the Minister of Munitions, to one Chaim Weizmann.

Weizmann was a Russian-born chemist who had studied in

Germany and Switzerland before coming to England to take up a post at the University of Manchester. At the time of his meeting with Lloyd George, Weizmann was involved in researching synthetic rubber; the English feared that the Germans would cut them off from their natural rubber suppliers in South America. Chemical analysis of rubber suggested that it could be made from a compound called isoprene, and Weizmann pursued this project fruitlessly, until fate intervened. After establishing himself in Manchester, Weizmann brought his fiancée, Vera Khatzmann, over from Europe and married her. As luck would have it, Vera's sister was married to a French scientist who worked at the Pasteur Institute in Paris. On a visit to his brother-in-law, Weizmann got a chance to look around the institute, and he discovered something remarkable. French scientists were using bacteria to convert various carbohydrate mixtures to simple chemicals through the process of fermentation. Weizmann now wondered whether some specific bacterium might produce the elusive object of his search, isoprene, which he could then use to make rubber.

It was a good idea, but it didn't work. No microbe that Weizmann tried yielded isoprene. But *Clostridium acetobutylium* did convert starch into a mixture of ethanol, acetone, and butanol, a blend that did not particularly interest Weizmann. It certainly interested Lloyd George, however; he heard the whole account from Weizmann himself. Here, pehaps, was a way to produce the acetone that they sorely needed for the manufacture of cordite. Weizmann was asked to scale up his experimental process, and within a short time he'd converted a gin distillery into a factory to make his mixture. He easily separated the acetone through distillation, and soon mass production was under way. There was no need for butanol, and huge stocks built up. But after the war the

automobile industry, which was growing by leaps and bounds, needed a car-paint solvent. Butanol was ideal.

The English discovered that the most readily available raw material for the fermentation needed to produce acetone was corn mash. When their own supplies were inadequate, they imported the mash from Canada. But, under threat from German U-boats, they realized that they had to find an alternate source of raw material in Britain, and that's when horse chestnuts entered the picture. They were plentiful, and they were a good source of starch. The trick was to collect them, so the government appealed to the nation's schoolchildren. They also set up a plant for converting horse chestnuts to acetone in King's Lynn, Norfolk, although for a time they kept the location a secret. Schools would send the chestnuts to government offices in London, and from these collection points the nuts were shipped to the plant. Unfortunately the collection system didn't work very well. Rotting chestnuts piled up at many a railway station. Still, whether from corn mash or chestnuts, Britain manufactured enough acetone to meet the Allies' demands.

They shipped the precious acetone to the Gretna explosives factory located in Scotland, outside of German bomber range.

Here the nitrocellulose and nitroglycerin were manufactured — it was a dangerous business. Workers were forbidden to bring hairpins and needles into the factory plant, and they had to wear rubber shoes — these were precautions against sparks. They mixed the guncotton and nitroglycerin by hand to make "the devil's porridge." Sulfuric acid and nitric acid fumes permeated the air, and the workers, mostly women, were speckled with burns from acid splashes, but they produced one thousand tons of cordite per week. Their efforts were commemorated in a popular song, which included these lyrics: "Give honor to the Gretna girls, Give honor where honor is due, Don't forget the Gretna girls, Who are doing their duty for you."

Lloyd George didn't forget the Gretna girls' contribution; nor did he forget Weizmann's. When he became prime minister, he asked Weizmann how he would like to be rewarded. Dr. Weizmann was an avid Zionist, and so he asked for British help in establishing a homeland for his people. Britain complied with the Balfour Declaration in 1917, establishing Palestine as a national homeland for the Jews. Weizmann eventually became the first president of Israel. And that is how acetone came to play a large role in forming world history. Think of that the next time you use it to remove nail polish.

From Obscure Physics to Practical Medicine

The January and February 1946 issues of the scientific journal *Physical Reviews* featured two independent research reports by Felix Bloch of Stanford and Edward Purcell of Harvard. The articles, "The Nuclear Induction Experiment" and "Resonance Absorption by Nuclear Magnetic Moments in a Solid," weren't bedtime reading. They described work that would seem to be of interest only to other physicists working in a specialized area. Many people likely questioned the wisdom of throwing away money on such esoteric research. But they would have been wrong. The work of these theoretical physicists resulted in a technique that revolutionized the fields of chemistry and medicine, two fields that lay well outside Bloch and Purcell's area of interest.

Both scientists were interested in measuring the strength of the natural, minute magnetic fields created by atoms of hydrogen. They devised experiments in which a sample of a compound containing hydrogen was placed between the poles of a large magnet, and they discovered that the poles of the small hydrogen "magnets" had a tendency to be attracted to the opposite poles of the external magnet. Then, most importantly, they found that they could force the tiny hydrogen magnets into a less desirable orientation by applying energy in the form of radio waves.

Interesting, but the question this prompts is, so what? The answer lies in how the results of that experiment evolved. Before long other scientists had taken Bloch and Purcell's findings and built on them. They discovered that not all of the hydrogens in a sample absorbed exactly the same amount of energy. The amount of energy absorbed was governed by the chemical environment of the hydrogens. From this information scientists were able to discern molecular structure, since most atoms in molecules reside in permanently different environments. Within a few short years, scientists around the world were using the new technique of nuclear magnetic resonance, or NMR, to identify the chemical structure of unknown compounds. Nobody was surprised when, in 1952, Bloch and Purcell were awarded the Nobel Prize for chemistry in recognition of their pioneering work. Today there is virtually no organic chemistry laboratory in the world that does not make extensive use of NMR.

In the early 1970s a group of scientists who were pushing the NMR envelope had a bright idea. They'd had enough of placing sample tubes in a magnetic field. Now it was time to experiment with living tissue. Why? Simply because it had never been done before. So they placed a live mouse between the poles of a large magnet. The procedure was totally painless for the mouse and extremely interesting for the researchers. The hydrogen nuclei in the various molecules that made up the body of the mouse aligned with the external field, as the scientists had expected, and those nuclei could be knocked out of alignment with low energy radio waves. Furthermore, when the energy source was turned off, the nuclei would release the energy they had absorbed. Some nuclei would release the energy quickly, others more slowly. Incredibly, the time taken by the various nuclei to "relax" was related to the health status of the tissue in which they were located. Within a short time scientists had designed a computer-linked instrument capable of producing cross-sectional views of body parts, and by the late 1970s they had built an instrument that could produce images of the entire human body. Nuclear magnetic resonance imaging was born.

The technique was wonderful. It was not invasive, and it used no dangerous radiation. It showed great promise as a tool for diagnosing conditions ranging from joint problems and cancer to strokes and spinal cord injuries. The problem, however, was the name. Patients were scared of the term "nuclear," which conjured up images of radioactive waste, atomic bombs, and things glowing in the dark. But the word "nuclear" doesn't necessarily have anything to do with radioactivity; it's just an adjective that describes any effect or property that pertains to the nucleus of an atom. Such as a magnetic field. The name problem was finally cleared up, not by educating the public, but by catering to the irrational fears: "nuclear" was dropped, and the technique was thereafter referred to as magnetic resonance imaging, or MRI.

Today magnetic resonance imagers are among the most important tools for probing the human body in a noninvasive fashion. Soft tissues, which defy X-ray analysis, reveal their secrets to MRI machines. The risks seem minimal. Patients with metals in their bodies, such as aneurysm clips, are not MRI candidates because the strong magnetic field can cause the metal parts to shift position. Sometimes patients complain that they feel claustrophobic inside the large magnet, but researchers have found that in many cases anxiety is alleviated by scenting the air with vanilla or fresh green apple fragrances. The most significant problem associated with MRI is a financial one. With our underfunded medical systems, we cannot make full use of this technique, which is poised to become the most important diagnostic advance in medical history.

Heart disease is our number-one killer, and MRI has a huge potential for detecting blockages in coronary arteries; and, unlike angiograms, it requires no physical intervention. It can identify the composition of a deposit, or plaque, in terms of fat, calcium, and fibrous muscle cells, and it can determine whether that deposit is likely to break open and trigger the formation of a blood clot that can cause a heart attack or stroke. On the basis of this, a physician may then decide to treat a patient at risk with drugs belonging to the "statin" family, which, at least in animals, have been shown to transform dangerous plaque into harmless deposits.

You'll soon be hearing a great deal about MRI-guided surgery: surgeons operate with nonmagnetic instruments while they and their patients are inside the magnetic field. The borders of tumors are often difficult to detect visually, and in many cases surgeons leave parts of the tumor behind. In MRI-guided surgery, tumors are clearly delineated. But, obviously, you can't buy such a setup with pocket change.

One of life's most alluring mysteries may also succumb to the power of MRI. At the Albert Einstein College of Medicine in New York, researchers have recruited people who profess to be in love and subjected them to MRI scans in an attempt to study the effects of Cupid's arrow on the brain. This is actually serious research. It is aimed at discovering why love sometimes degenerates into jealousy, which is often associated with battering. Does this mean that a scientific answer to the age-old question, "Do you really love me?" is just around the corner? At least one of the subjects in the study wasn't taking any chances. He really loved his girlfriend, he insisted, but didn't want her around when his brain was being imaged, just in case his love didn't show up.

The birth of an obscure theory, the molecular-structure determination, the leap to medical technology, the revelation of cerebral mysteries: that's the history of magnetic resonance. Could a better case be made for funding research? I think not. You can never tell where it will lead — like the destruction of the myth that we use only ten percent of our brains. This claim has been made by some psychics, who suggest that we would be capable of incredible feats if we just learned to use our brains to full capacity. MRI has shown that most of us use over ninety percent of our brains, although not necessarily all parts at the same time. But perhaps some people do use just ten percent of their brains — those people who believe in the ten-percent myth.

The Not-So-Miraculous Elixir

Little Joan had a sore throat that just wouldn't go away, so her mother took her to the doctor. "Say ahhhh," the doctor instructed. He quickly diagnosed a streptococcal infection. "Do you like raspberry syrup?" he asked. "Sure do!" And with that, mother and daughter were on their way to the local drugstore to fill a prescription for Elixir Sulfanilamide. The doctor did not expect to see this patient again for the same complaint. This was 1937, and, thanks to a miraculous new drug, streptococcal infections were no longer the scourge they had once been. Sulfanilamide, which had been introduced a couple of years earlier, had proven its value not only against strep throat but also against gonorrhea and meningitis. On many occasions the doctor had prescribed sulfanilamide in the form of a powder or a pill, always with excellent results. But children didn't like taking pills, so he was pleased that at last someone had come up with a tasty sweet syrup that would make the medicine go down easily. Elixir Sulfanilamide did just that. It also killed little Joan.

Death had not come quickly. For days she'd endured vomiting, excruciating abdominal pain, and convulsions. When Joan's suffering finally ended, her distraught mother wrote to President Roosevelt: "Even the memory of my little girl is mixed with sorrow, for we can see her little body tossing to and fro and hear that little voice screaming with pain. It is my plea that you take steps to prevent the sale of such drugs that will take little lives and leave such suffering behind and such a bleak outlook on the future as I have tonight." President Roosevelt took action, for this was not the first such case that had been brought to his attention. The Food and Drug Administration had informed him that over one hundred deaths had been attributed to Elixir Sulfanilamide. One single physician had reported the deaths of six of his patients, including his best friend.

In a heart-wrenching letter to the FDA, that doctor described his mental and spiritual agony over having unknowingly prescribed a deadly medication.

How could such a dangerous drug have made it to the marketplace? Easily. Incredible as it seems to us now, at the time manufacturers could sell drugs without testing them at all. Of course, it wasn't in a drug company's interest to sell harmful substances, but drug manufacturers weren't bound by any legal safety requirements. This had to change, Roosevelt decreed, and within a year the 1938 Food and Drugs Act became law. Manufacturers would now have to seek government approval to sell a drug; and to get such approval they would have to prove that their product was safe. (They still did not have to show that the drug was effective — that requirement came only in 1962, on the heels of the thalidomide tragedy.)

The way Elixir Sulfanilamide was brought to market illustrates just how lax the prevailing state of affairs was. Although sulfanilamide pills were already widely used, patients in the southern United States tended to reject the tablets because they had long been accustomed to taking their medications in the form of sweet-tasting solutions called elixirs. To accommodate them, the S.E. Massengill Company of Tennessee set about developing a liquid form of sulfanilamide. Harold Watkins, one of the company's chemists, undertook the project.

Watkins had trouble dissolving the drug in both water and alcohol, so he tried a number of common laboratory solvents. A sweet-tasting liquid called diethylene glycol fit the bill. Although the only testing he did was to taste the occasional sample himself, Watkins developed a method for making raspberry-flavored sulfanilamide elixir on a grand scale. Within a couple of months doctors nationwide were prescribing the formula, mostly for children suffering from infections. Tragically, the treatment turned out to be more deadly than the disease for 107 people.

It seems incredible now that no one bothered to check the toxicity of the solvent Watkins used. Had they bothered to peruse the scientific literature, they would have discovered that diethylene glycol is metabolized in the body to oxalic acid, a potent kidney toxin that can kill. Simple animal experiments would have confirmed the risk.

But, oblivious to all of this, Massengill manufactured and distributed 240 gallons of Elixir Sulfanilamide in September of 1937. By October reports of deaths had come in, but the company refused to shoulder any of the blame. Massengill spokesmen maintained that they had broken no laws. Furthermore, they refused to divulge the contents of the formula because it was a "trade secret." Finally, under pressure from the American Medical Association, the company revealed that the elixir contained diethylene glycol. The nature and scope of the problem was now clear, and the American Medical Association, the FDA, and the media joined forces in a massive effort to track down as much of the elixir as possible. They were astonishingly successful: 234 of the 240 gallons were recovered.

The only charge that could be brought against Massengill was one of false labeling — according to a 1906 law elixirs had to be made with alcohol. But the public outcry over the sulfanilamide tragedy, coupled with President Roosevelt's resolve, led directly to the passage of the Food, Drug and Cosmetic Act of 1938. At last legislation existed that would compel drug manufacturers to prove the safety of their products prior to marketing them.

Today the research behind pharmaceutical products is so extensive that it often takes manufacturers over ten years and two hundred million dollars to bring a new drug to market. We have learned our lesson. Deaths like the 107 associated with Elixir Sulfanilamide are unlikely to occur in our time. Actually, it was 108 deaths. While Massengill admitted no liability, the company did fire Harold Watkins anyway. Unable to live with his feelings of guilt, the chemist shot himself through the heart. He was the tainted elixir's last victim.

SILLY STUFF

BENDING SPOONS OR BENDING MINDS?

You should go to a magic convention at least once in your life. You'll be fooled and entertained as coins vanish, selected cards rise out of the deck, and ten-dollar bills float in the air before your eyes. But, most importantly, you'll never look at the world the same way again. Frankly, I can't think of a better way to develop critical thinking than getting fooled by the honest charlatans at a magic convention.

What is an "honest charlatan?" Let's start by defining "charlatan." Simply put, a charlatan is someone who pretends to have some power, skill, or knowledge that he or she does not actually possess. A charlatan's claims can range from an ability to remove tumors without making an incision to an aptitude for bending spoons with the mind. An "honest" charlatan can produce the same effects but will freely admit that it's all trickery. The dean of honest charlatans is The Amazing Randi, one of my idols. He is a world-famous magician, but, more significantly, he is also the superman of rational thought. He fights for truth, justice, and the scientific way. What a delight it was for me to finally meet The Amazing One at a magic convention in Montreal. For two hours we chatted about the current belief in various types of silliness and the importance of exposing fraud wherever it exists.

Randi has built a formidable career on such exposures. And he has put his money — at the moment it's over one million dollars — where his mouth is. Anyone who can evoke a paranormal phenomenon under controlled conditions can claim the cash. Determine the contents of a sealed envelope through telepathy, move an object by "psychokinesis," or bend a spoon with your mental power, and the money's yours. While Randi has tested a host of challengers, no one has walked away with the prize. Uri Geller hasn't even tried. Oh yes, Uri Geller. It is virtually impossible to discuss Randi without mentioning Geller, the psychic superstar who, for nearly three decades, has been bending spoons and minds for a living.

Geller, a former Israeli, is a magician with a charming manner. He claims to have abilities that he himself doesn't even understand. He gently rubs keys and they bend; moving his hands above sealed cannisters, he identifies the ones containing water. Yet, strangely, he cannot do these things when Randi is around. When Geller first came to the United States he appeared with Johnny Carson on *The Tonight Show*. Everyone was eager to see him, because he'd already captivated huge live audiences with his psychic feats. Now millions of TV viewers would finally get a chance to see the phenomena that science could not explain.

The appearance was a fiasco. Geller was unable to accomplish anything. He didn't feel right, he said — the energy just wasn't there that night. But it was quite apparent that Geller's psychic powers had failed him even earlier, otherwise he would have known that Johnny Carson was an amateur magician and that the show's producers had found out from Randi how "psychic" feats could be enacted with magicians' tricks. Geller couldn't bend the spoons supplied by the show; and he couldn't determine which sealed film cannister contained water, because Randi had advised the producers to attach the cannisters firmly to the table. Geller's trick of shaking the table imperceptibly to see which cannister moved was therefore foiled. Only Geller was visibly shaken.

Strangely, the psychic flop did not destroy Geller's career. His next TV appearance was on *Donahue*, and this time everything worked. Proof, Geller declared, that he was not a magician. He was for real — if he was doing tricks, then they would work every time. Eventually, articles and books appeared explaining how Geller performed his stunts, and North American interest in the spoon bender declined. He moved to England, and these days he helps companies find gold or oil through psychic map reading and markets Uri Geller's Mind Power Kit with crystal quartz for psychic healing.

Believing in psychic spoon bending may seem harmless enough, but it's not. If you think a person can bend metal with thought processes, then you're also capable of believing that a person can remove a tumor from your body without making an incision. Practitioners of "psychic surgery" exploit standard sleight-of-hand tricks, and the procedure looks very impressive, but what happens to cancer patients who are taken in by the ruse? Might they be foregoing some effective therapy?

Randi's purpose is not to debunk uncritically. He evaluates claims fairly. Like any scientist, he would be thrilled to see our scientific sphere expanded. How exciting it would be if we could send messages to each other telepathically, if we really were being visited by aliens, if we could bend metal with our minds. I could have chatted with Randi on and on, but two hours had flown by and I couldn't monopolize The Amazing One any longer. Anyway, I wanted to go and check the magic dealers. My psychic key bender had worn out, so I went to get a new one. While I was at it, I found a new spoon effect. "Spoon

Spinner" makes it possible for a magician to change an ordinary teaspoon with his "aura" and use it to pick up another spoon. I had to have it.

I could hardly wait to get home and demonstrate my acquisition to my family, but when I took the spoons out of my pocket one of them was bent. Could it be that I have some unrecognized psychic power? Maybe. Or maybe I just sat on the spoon. I wondered what Uri Geller would have said, and, as luck would have it, I got a chance to ask him. To my surprise, the famed "paranormalist" had agreed to a phone interview with me on the radio. Why was I surprised? Simply because Geller normally crops up on national TV shows like *Oprah* and *The View*, but I guess people with a product to promote become more willing to try new forums. After all, there's money to be made. Geller's product was a book, a rather interesting one, titled *Uri Geller's Mind Power*. We never got around to talking about it, however. A few minutes after our conversation began Geller hung up on me.

I must confess that I've been fascinated by Uri Geller for some time. My fascination dates back to the early 1970s, when I was a chemistry student at McGill and hadn't even heard of The Amazing Randi yet. I received a letter from an uncle in

Israel describing a nightclub performance he had seen. An entertainer had wowed the crowd with a number of mind-reading effects and then called a young lady to the stage. He placed a piece of aluminum foil in her hand and said that he would heat it up with his mental power alone. Sure enough, within seconds the volunteer let out a scream and dropped the aluminum like a hot potato. My uncle was impressed, but he did have a few thoughts on the matter — he was a chemist, and chemists are intrigued by matter and the changes it undergoes. He wanted to know whether I had come across anything in the course of my studies that would illuminate the chemical reaction behind a feat like the one he'd witnessed. As it happened, I had — but not in my chemical studies.

As a youngster I had become interested in magic as a hobby. We're not talking about casting spells here; we're talking about pulling rabbits out of hats, making coins vanish, and, yes, heating aluminum foil. There's an old trick in which the magician secretly slips a bit of mercuric chloride into a piece of folded aluminum before placing it in a spectator's hand. As if by magic, the aluminum rapidly becomes too hot to handle. Of course, this is due to an exothermic chemical reaction, not magic. I used to do this trick until I started to study chemistry, at which point I pulled it from my repertoire permanently. Mercuric chloride is potentially toxic, and it isn't wise to handle it. When I communicated this bit of chemical trivia to my uncle, he agreed that it could explain how the entertainer had pulled off that stage stunt. Still, my uncle said, the man had performed the trick well, and he remained impressed with him. He told me that if I ever had the chance to see that entertainer, I should go. The man's name was Uri Geller, and that was the first time I heard the name that would eventually become world famous.

You can imagine, then, that when Uri Geller came to Montreal in the early 1970s I was there in the ballroom of the old Mount Royal Hotel, ready to take in what I thought would be some sort of magical entertainment. But there was no show — not in the traditional sense, anyway. Geller, in his usual charming fashion, told us that as a child he'd realized that he possessed the ability to deform metal objects simply by concentrating on them. He didn't understand why he had this power, but he had it. As I recall, the evening consisted mostly of Geller telling us what he had done and what he would do, but there wasn't very much doing. He did display a selection of spoons and keys, some of which he'd collected from members of the audience. Then there was a lot of talk, a lot of metal stroking, and, suddenly, a cry of surprise from Geller himself. He seemed more shocked than anyone as he triumphantly held a bent spoon aloft. The audience oohed and aahed.

I wasn't all that impressed, because although Geller had kept repeating, "it's bending, it's bending" as he was handling the spoon, I hadn't detected any movement. What I did see was an unbent spoon at the beginning of the demonstration and a bent one at the end. In the interim there had been much commotion and much misdirection, a practice I had become familiar with through my own attempts at prestidigitation. Misdirection is manipulating an audience to look at the wrong place at the right time. By the time Geller got to twisting keys and started stopping watches, I was paying close attention, and I began to get a handle on what was happening. I had come to see Geller expecting to witness some clever deceptions, but not of this variety. I had not anticipated being told that the effects we were seeing were accomplished by purely psychic means, devoid of trickery.

After that I followed Geller's career closely. I watched his television appearances, I read his books, and I read about him. Randi's outstanding work *The Truth About Uri Geller* filled in the blanks for me and explained how the psychic could have accomplished all of his "miracles" by means of various tried-and-tested conjuring tricks. As I became more and more involved with science, I became more and more bothered by Geller's claim that he had special powers. Energy can, of course, be converted from one form to the other, but it cannot be created or destroyed. It takes a lot of energy to bend a key or a spoon. Where was this energy coming from? Geller attempts to address this question in *Mind Power* by telling us that coal has hidden energy that is only released when we light it; similarly, he asserts, thoughts and feelings can also generate energy. Yes, there is some truth to that, but here we are not talking about a combustion process. We can only measure brain waves with sophisticated instruments, and the electrical energy associated with these waves is minute in comparison to the energy required to bend a spoon! We can't even bend a spoon by putting it next to high-voltage power lines!

So you can see why I was eager to interview Uri Geller, especially since he seemed to have reinvented himself as a "healing facilitator." I wanted to give him some credit for getting away from spoon mangling and concentrating instead on the potential of mind power as a healing technique. And I wanted to commend him for the charitable work he'd done for hospitals. But I also wanted to ask him why he had never tried to claim the James Randi Educational Foundation's million-dollar prize. If he'd just perform a psychic feat under controlled conditions, then he'd have a million dollars to help sick kids with. Wasn't it irresponsible of him not to take advantage of Randi's offer? Wouldn't it be better than giving the children "Uri bears," which have crystals (energized by Geller himself) hanging around their necks? Beyond this there was so much more I wanted to ask. I was curious about his sighting of a UFO in the Israeli desert, about the women in his audiences who experience spontaneous erogenous satisfaction, and about how he had managed to become an expert on electric fields, cosmic radiation, hormones, drugs, and pollution.

My questions, however, would remain unanswered. After I introduced myself and recounted the aluminum story, he stated that he'd never performed a stunt like that. Perhaps my uncle had seen a Uri Geller impersonator. I doubt it: I found a 1974 Israeli newspaper article that mentions Geller's use of the aluminum trick. Anyway, I shifted the topic to *Mind Power* and began to ask a question about Iscador, a mistletoe extract that Geller claims in the book has an anticancer effect. Before I could complete my question the great Uri Geller, the man who reads minds, the man who terrorizes cutlery, delivered his ultimate comeback. He hung up. Just as Randi told me he would at any hint of confrontation. The Amazing Randi, it seems, is the one with the real psychic powers.

Down on the Biodynamic Farm

I had never eaten a biodynamically grown tomato. In fact, I had never even heard of such a thing, but the saleswoman in the Vancouver health food store assured me that this tomato was not only "free of chemicals" but had also grown in harmony with the moon and the planets. I learned long ago that it is fruitless for me to get into discussions with the scientifically challenged, so I bit my tongue and, sensing a good story, invested a small fortune in the biodynamic tomato. Storywise, it turned out to be a good investment. I never imagined that a little bit of research into biodynamic farming would lead me to the strange world of anthroposophy and its mystical founder, Rudolf Steiner.

Steiner was born in 1861 in a part of the Austro-Hungarian Empire that would later become Yugoslavia. He was an architect by training, and some of his unusual free-flow structures still stand in Germany and Scandinavia. But it was in a very different area that Steiner achieved fame — or notoriety, depending on one's views. In 1899 Steiner embraced a fledgling religion called theosophy. The Ukrainian psychic Helena Petrovna Blavatsky was theosophy's founder, and she based the belief system on a curious blend of astrology, spiritualism, and Eastern mysticism. Blavatsky claimed to be in contact with spirits who would send her written messages; during her séances these notes would float mysteriously down from the ceiling. On numerous occasions Blavatsky was caught with conjuring equipment and her spirit manifestations were exposed.

Steiner eventually split with Blavatsky. He founded his own peculiar religion, which he called "anthroposophy." The name derives from the Greek words for "man" and "wisdom," but the general tenets of anthroposophy are virtually impossible to describe. Steiner maintained that he was clairvoyant, and in his writings he stressed the importance of "bringing oneself into harmony with the divine creative force" and "accessing a cosmic reservoir in the astral plane where every thought or action that has ever occurred is recorded."

Steiner also emphasized the oneness of the spiritual and material worlds and the importance of balancing cosmic forces. Such lingo is not unusual in metaphysical movements, but anthroposophy goes beyond mystical exhortations. Steiner had some, let us say, "unusual" ideas about how the spiritual and the material worlds are linked. The soil, he said, is a living organism with a life force that we must replenish constantly. Agricultural chemicals destroy the soil, so we have to nourish it with special composts and sprays in order to make it more "spiritually balanced."

So far this just sounds like an odd rationale for applying fertilizer, but Steiner's methods don't actually make use of any scientific principles. "Biodynamic farming," a term he coined, requires the use of certain "biodynamic solutions." The biodynamic farmer prepares one of these solutions by packing cow manure into a cow's horn and burying it. After several months the farmer digs up the horn and, in a ritualistic fashion, mixes the powder inside it with twelve gallons of rainwater. The farmer stirs the solution for twenty seconds in one direction and then reverses the direction for twenty seconds to "bring the cosmic life forces of the earth and the universe into harmony." This mystical mix, insisted Steiner, will "rejuvenate" four acres of soil. Chemically speaking, by this time the manure would be so diluted that it couldn't possibly have any effect. But, then again, anthroposophists don't believe that the fertility of the soil has anything to do with such mundane things as chemicals.

Once the farmer has prepared the soil, he or she can plant the crops — bearing in mind the position of the moon and the planets. Steiner thought that the gravitational pull of the moon was critical to the nurturing of plants and that new crops should be planted two days before a full moon, regardless of the fact that the phases of the moon are not linked to that body's gravitational effect on the earth. Planetary positions, Steiner claimed, also determine when different parts of a plant grow. Farmers should tend lettuce or spinach on "leaf" days and potatoes on "root" days. Steiner was also convinced that the chirping of birds and the flapping of their wings influenced plant growth. The right frequencies spurred the plant's development, he insisted.

There are still farmers who practice biodynamic farming in several countries; some of them even use a concert of birdsongs. And their claims are becoming more and more fantastic. A couple of years ago the Biodynamic Farming and Gardening Association of New Zealand offered to come to the aid of the government. Possums, which the New Zealand government had imported from Australia with the aim of establishing a fur industry, were destroying forests by eating everything in sight. Neither trapping nor poisoning had made a dent in their numbers, but the biodynamic farmers had a solution. Burn possum testicles, they advised, and blend the ash with sand to make "possum pepper." Next spread a highly diluted extract of this concoction around the possums' habitat. The possums would flee in terror after getting a mere whiff of this emasculating stuff. The desperate New Zealand Forest Institute actually carried out a double blind, placebo controlled study of this bizarre technique on both wild and penned possums. The biodynamic materials had absolutely no effect: the possums did not even play possum.

Anthroposophists also take a unique approach to medicine. The body, they say, has three poles — "cool," "warm," and "balancing." Illness arises from a disharmony of these poles, and we can restore the harmony with a variety of animal, mineral, and plant substances. We must take the time of day and planet constellations into consideration as we prepare these remedies, and we must never work on them between noon and three o'clock in the afternoon because this is the "least alive" time of day. Anthroposophists frown upon vaccinations, but they approve of color therapy and a mistletoe preparation invented by Steiner. Sauerkraut they regard as a special food; we require it for the health of our digestive tracts.

To a scientist, this is pretty silly stuff. And Steiner himself admitted it: "I know perfectly well that all of this may seem utterly mad," he once said. "I only ask that you remember how many things have seemed utterly mad which have nonetheless been introduced a few years later." An interesting notion, but quite misleading. The fact is that most ideas that seem utterly mad are utterly mad. But, at least in one instance, Rudolf Steiner's views transcended the usual metaphysical drivel. He states in one of his books that "If the blond and blue eyed people die out, the human race will become increasingly dense if men do not arrive at a form of intelligence that is independent of blondness." Scary stuff.

Anyway, I now know what biodynamic farming is all about. And what did my tomato taste like? Like any other tomato. Maybe the birds didn't sing to it enough, or maybe the planets were not properly aligned when it was planted. The saleswoman agreed that the tomato was not as tasty as usual. Perhaps the grower had not done everything required. Perhaps that grower had neglected to perform "earth acupuncture." Don't even ask.

The Goat Gland Doctor

The remarkable events I'm going to chronicle here would likely never have unfolded, in 1917, if young Dr. John Brinkley had not been hired as house doctor at the Swift meatpacking company, located in Kansas. He was dazzled by the vigorous mating activities of the goats destined for the slaughterhouse. A couple of years later, after Brinkley had gone into private practice in Milford, Kansas, a farmer named Stittsworth came to see him. Stittsworth complained of a sagging libido. Recalling the goats' frantic antics, the doctor semi-jokingly told his patient that what he needed was some goat glands. Stittsworth quickly responded, "So, Doc, put 'em in. Transplant 'em."

Most doctors would have ignored the bizarre request, but Brinkley was not like most doctors. In fact, he wasn't a doctor at all. Although he had spent three years at Bennet Medical College in Chicago, he'd never graduated. He called himself a doctor on the basis of a five-hundred-dollar diploma he had purchased from the Eclectic Medical University of Kansas City, Missouri. As absurd as it sounds, this piece of paper gave him the right to practice medicine in Arkansas, Kansas, and a few other states.

Buying a degree from a diploma mill was not out of character for Brinkley. He had worked as a snake-oil salesman in a road show, and then, with Chicago con man James Crawford, established Greenville Electro Medical Doctors. Under this name the pair injected people with colored distilled water for twenty-five dollars a shot. And that was big money in those days. Brinkley, therefore, had all he needed to capitalize on the farmer's idea of goat-gland transplants: he was unethical, he had a wobbly knowledge of medicine, he had witnessed the rambunctious behavior of goats. And he possessed one more thing: knowledge of experiments carried out in Europe beginning in the late 1800s.

Charles-Edouard Brown-Sequard, a noted French physiologist, had shocked the medical community by injecting himself with the crushed testicles of young dogs and guinea pigs. Afterwards he claimed that he had regained the physical stamina and intellectual vigor of his youth. Many men availed themselves of La Méthode Sequardienne, but once the placebo effect was filtered out little remained. In Vienna physiologist Eugen Steinach proposed that youthful vitality could be restored by increasing levels of testosterone. The easiest way to do this, Steinach said, was through vasectomy. Sperm production wasted testosterone, and if the channel leading from the testes to the ejaculatory duct were tied off, then blood levels of testosterone would rise. Brinkley may also have heard of the work of Serge Voronoff, a French doctor who was stirring up a storm of controversy with his experimental gland transplants. Voronoff had been a physician in the court of the King of Egypt, and there he had spent a great deal of time treating the court eunuchs, who suffered from a variety of illnesses. He hypothesized that maintaining active genital glands was the secret to health. As proof, he cited his experiments with an aging ram into which he had transplanted the testicles of a young lamb. The ram's wool got thicker, and his sexual vigor returned. Voronoff then went on to transplant bits of monkey testes into aging men; he claimed success, although he could offer no scientific validation of his claim. In America the stage was set for the meteoric rise of J.R. Brinkley.

Brinkley went to work, implanting a bit of goat gonad in Stittsworth's testicle. Within weeks the farmer was back to thank the doctor for giving him back his libido. And when his wife gave birth to a boy, whom they appropriately named Billy, Stittsworth spread the word about Brinkley. Soon Brinkley's business was booming. The testimonials poured in and so did the money. Brinkley was charging $750 per transplant, and he couldn't keep up with the demand. All men needed the Brinkley operation, he declared, but the procedure was most suited to the intelligent and least suited to the "stupid type." This, of course, ensured that few of his patients would admit that they had not benefited from the operation.

There were a few problems. Like when Brinkley decided to use angora goat testicles instead of those from the more common Toggenberg goat. Recipients of the angora testicles were decidedly unhappy — Brinkley himself noted that they reeked like a steamy barn in midsummer. But Brinkley's major problem was that as his fame increased so did the criticism leveled against him by the medical community. Morris Fishbein, editor of *The Journal of the American Medical Association*, called Brinkley a smooth-tongued charlatan and urged the authorities to revoke his right to practice. Brinkley's assertion that his procedure could cure conditions ranging from insanity and acne to influenza and high blood pressure amounted to quackery, Fishbein said. In response to this Brinkley called the American Medical Association a "meat-cutters union" and charged that

its members were jealous of him because they were losing business. He then went to California and performed a transplant on Harry Chandler, the owner of *The Los Angeles Times*; the satisfied Chandler rewarded Brinkley with lots of free publicity.

In California Brinkley also learned about the potential of radio. Returning home, in 1923, he started up the radio station KFKB with one thousand watts — an amazing number for the time — and broadcast music, his lectures on rejuvenation, political features, and the "Medical Question Box," during which Brinkley himself answered listeners' questions. It was perhaps radio's earliest advice show. But the advice Brinkley dispensed was ridiculous, and he usually gave listeners prescriptions to which he assigned a number. These they could fill at a local pharmacy; Brinkley had set up the National Dr. Brinkley Pharmaceutical Association in collusion with pharmacists who relished making lots of money selling water colored with indigo.

Kickbacks from this operation and revenues from the transplant surgeries made Brinkley an immensely wealthy man. For five thousand dollars he would even implant genuine human glands, which he obtained from prisoners on death row. He had mansions, a fleet of Cadillacs, airplanes, and yachts. What he did not have was scientific respect. The American Medical Association finally prevailed upon the Kansas Board of Medical Registration to revoke Brinkley's license on the grounds of immorality and unprofessional conduct, and the Federal Radio Commission shut down KFKB for promoting fraud. Still Brinkley did not capitulate. He claimed he was being crucified by the authorities and kept his hospital going by hiring licensed physicians to work there. He also purchased radio station XERA in Mexico and began beaming his message into the United States with the power of one hundred thousand watts.

The "doctor" then decided that the only way to get his license back was to become governor. So in 1930 he organized a massive write-in campaign, and he almost won. By insisting that he was being persecuted by elitist doctors and politicians he won the support of ordinary citizens; and his promise to build free clinics and cure virtually all diseases boosted his appeal even further. But Brinkley couldn't even cure himself. The Milford Messiah — as he was sometimes called — the man who had performed over sixteen thousand goat testicle transplants, the man who appropriately wore a goatee all his life, developed a blood clot, and doctors had to amputate his leg. Till the very end, Brinkley's scheming mind remained active. Confined to bed, he decided to study for the ministry and had visions of becoming a big-time preacher. He never made it. His last words were reported to have been, "If Dr. Fishbein goes to heaven, I want to go the other way." If there is any justice in the world, he did.

A Writer and a Magician Among the Spirits

Spirits are everywhere. And not just on Halloween. Furthermore, they're not silent, they're chattering away — but not directly to us, of course. Ghostly conversation, it seems, is only possible through a spiritual medium. Their numbers are growing these days, and their services don't come cheap. George Anderson used to be a Long Island telephone operator, but now he charges four hundred dollars for a connection and he doesn't even need a phone. The current dean of mediums is New Yorker James van Praagh, who plays to packed houses for up to forty-five dollars a head and does private readings for celebrities such as Cher. (Apparently, Sonny liked his funeral but is not happy with his burial site.)

While a belief in the spirit world is part and parcel of many religions, spiritualism, specifically, is a philosophy built on communicating with the dearly departed. Spiritualism dates back to 1848, when Katherine and Margaret Fox began fooling the gullible by contacting the other side in their own unique fashion. These two young sisters discovered that fame and fortune lay in "spirit rappings." They astounded their family members, then their neighbors, and finally huge audiences by subjecting them to strange cracking sounds, which, they insisted, emanated from the spirit world. The noises occurred while Katherine and Margaret sat motionless, apparently in a semitrance, concentrating on connecting with the other side. They were concentrating, all right, but not on spirits. They were concentrating on their toes. The two had an amazing talent: they could crack these digits to produce eerie, haunting sounds. Katherine and Margaret made a career out of exhibiting their strange skill, and they somehow remained famous as mediums even after publicly owning up to their bit of mischief. In 1881, on stage at the New York Academy of Music, before a panel of three physicians, Margaret Fox took off her shoe and demonstrated her toe-cracking talent to a stunned gathering. In spite of this, supporters refused to believe that the original rappings were not produced by the spirits. The antispiritualists, they insisted, had coerced Margaret into this absurd demonstration.

The Fox sisters got people thinking about spirits, but it was the Davenport brothers of the United States who furnished physical evidence that such forces did, indeed, exist. Capitalizing on the publicity generated by the Foxes, the Davenports devised an act during which selected members of the audience tied them securely to chairs inside a large "spirit cabinet" furnished with a variety of props. Soon music began to play, bells rang, and trumpets appeared through openings in the cabinet; yet when the doors were opened at intervals in the "performance," the brothers appeared still to be bound to their chairs. Were all the wonderful manifestations attributable to spirit power? Actually, they were attributable to a few clever rope tricks and some great showmanship. Penn and Teller, modern magicians par excellence, have re-created the Davenport's act, and they continue to confound audiences with it. It's a very spirited presentation, but the duo gives no impression that spirits are involved. Back in the 1800s, however, audiences could not have imagined how else one could explain such extraordinary occurrences.

Sir Arthur Conan Doyle, of all people, was completely taken in. The creator of Sherlock Holmes, the most logical and scientific of fictional detectives, was a firm believer in spiritualism, in spite of the fact that he was a physician by training and took great pride in his powers of observation. One day, while on rounds at an Edinburgh hospital, he stopped to examine a sick baby. He then told the child's mother that she must stop painting the baby's crib. When the startled lady asked him how on earth he knew that she'd done this, Conan Doyle explained that the boy was pale, listless, had no strength in his wrists, but was well fed. These symptoms suggested lead poisoning, and the diagnosis was confirmed in Conan Doyle's mind by the specks of white paint he noticed on the mother's hand. In those days white paint contained a large amount of lead, and Sir Arthur knew that children had a tendency to chew on the rails of their cribs. He had made a brilliant deduction based on some simple observations. Yet this man was completely bamboozled by the shenanigans of the Davenports.

Conan Doyle's lack of skepticism is surprising, to say the least. Especially since he was, at least for a while, a close friend of the great Harry Houdini, the scourge of fake mediums. Houdini was unable to convince Conan Doyle that the Davenports's effects were achieved by trickery. Furthermore, Sir Arthur believed that Houdini himself had unacknowledged paranormal powers — after all, how could he possibly carry out his fantastic escapes if he didn't dematerialize his body? This presented Houdini with a frustrating conundrum; the only way to persuade his friend that he was wrong would be to reveal his precious professional secrets.

The two eventually had a falling out, after Houdini agreed to attend a séance during which Conan Doyle's wife planned to contact Harry's dead mother, Mrs. Weiss. When Lady Doyle had, apparently, established a connection, she became the conduit for Mrs. Weiss's written messages. Houdini was disturbed that the first message Lady Doyle wrote out began with the sign of the cross. His mother, the wife of a rabbi, would never have used this symbol. Also, the message was in perfect English, a language that Mrs. Weiss had never mastered — she'd always spoken Yiddish to her son. (Who knows, perhaps the lady had been taking English lessons on the other side.) In any case, the pragmatic Houdini was no longer able to stomach Conan

Doyle's psychic foibles, and he dissolved their long-standing friendship.

It's interesting to note that the friendship had survived the celebrated Cottingley fairy episode. Conan Doyle had become convinced that tiny fairies, just like the ones in children's tales, really existed — he'd seen the photographs, and they "couldn't possibly be faked." But, of course, they were, and not even particularly well. One of the photos featured ten-year-old Francis Griffiths cavorting with a fairy in a glen in the English village of Cottingley. The man who masterminded the Sherlock Holmes stories was fooled by a cardboard cutout hanging on a thread. The fact that the fairy's fluttering wings were not blurred but a waterfall in the background was meant nothing to him; Conan Doyle just couldn't imagine that a couple of working class girls could fool an aristocrat like himself. There are none so blind as those who will not see.

Sir Arthur Conan Doyle went to his grave believing in fairies and spirits. One wonders what he would have said if he had known that *Princess Mary's Gift Book*, published two years prior to the Cottingley fairy episode, featured fairy pictures identical to the photos that had so enthralled him. Perhaps we should enlist James van Praagh to find out — after all, he talks to the dead. But I'm not sure we would get an answer. For, as Houdini himself said, anyone can talk to the dead. The problem is that the dead don't talk back.

In a Lather over Toiletries

"Are you poisoning your family?" "Is your bathroom killing you?" "Yes!" proclaim numerous Web sites, where there's much discussion about the culprits, sodium laureth sulfate (SLES) and sodium lauryl sulfate (SLS). What are these molecules, and why are people saying such horrible things about them? The what is easy to answer; the why is somewhat more difficult. Both of these substances are synthetic detergents, or syndets, widely found in cleaning agents ranging from shampoos to dishwashing products. They were originally developed by chemists to eliminate the formation of soap scum. Soaps, unlike detergents, react with dissolved minerals in water to form an insoluble precipitate, which often manifests itself as the classic bathtub ring.

The original discovery was accidental. In 1831, in France, Edmond Fremy treated olive oil with concentrated sulfuric acid and noted that after neutralizing the mix with sodium hydroxide he was able to isolate a product that lathered in water and cleaned like soap. It did not, however, react with dissolved magnesium or calcium. By the early 1920s chemists had created a number of detergents by reacting various fat derivatives with sulfuric acid. One of these was lauryl alcohol, found in coconuts. Sodium lauryl sulfate turned out to be an excellent foaming and cleaning agent. It's a long molecule, one end of which dissolves in fat and the other in water. In this way it forms a link between fatty deposits and water, allowing the fat to emulsify and rinse away. But SLS did present one problem: it was not soluble in cold water. This was not a concern when it came to laundry detergents, but it did mean that most shampoos, which needed to be clear at room temperature, had to have a limited sodium lauryl sulfate content. So the chemists kept tinkering, and they finally developed sodium laureth sulfate, a substance that was soluble and ideal for see-through shampoos.

As cleaning compounds, both SLS and SLES turned out to be cheap and effective, so they quickly found their way into a wide variety of shampoos, bubble baths, and even toothpastes. They have also found their way onto numerous Web sites.

Judging by the rhetoric used on those sites, one might think that these common detergent ingredients are a plague that soap manufacturers have unleashed upon humanity. They'll make your hair fall out, give you cataracts, accumulate in your organs, and, of course, they cause cancer. And, as if that weren't enough, the Web sites also maintain that auto mechanics use sodium lauryl sulfate to wash garage floors. This is undoubtedly true, but you could say the same thing about the water in shampoo: we use it to clean garage floors, and toilets as well.

What has inspired the outlandish accusations? Scientific ignorance is at least partially responsible. There is much discussion on the Web sites in question about the degree to which SLS and SLES are contaminated with carcinogenic nitrosamines. While it is true that trace amounts of nitrosamines have been found in some shampoos, they have nothing to do with sodium lauryl or laureth sulfate. Ethanolamine lauryl sulfates, substances found in a few products, accounted for the insignificant levels of contamination. The similarity in names, I guess, led the chemically illiterate to slur the much more common detergents.

Even more absurd is the allegation that sodium lauryl sulfate is carcinogenic because it contains dioxin. Here, panic arises from a simple spelling mistake. Dioxin is a notoriously dangerous chemical, but it has nothing to do with sodium lauryl sulfate. Dioxane, however, is a solvent used in SLS manufacture and may remain behind in trace amounts, but it is not hazardous. Perhaps someone just thought that dioxane was dioxin and felt that it was his or her duty to alert the world. (Microsoft Word's spell check also thinks that "dioxane" is "dioxin" misspelled, and it prompted me to replace one with the other. Does Bill Gates need a chemistry lesson?) But there may be yet another reason for the attack on SLS — a more nefarious one. The propaganda seems to be originating from

Web sites maintained by independent distributors of so-called natural cosmetics, people who hype the fact that their products do not contain "harsh detergents." Could there be a marketing motive here?

The anti-SLS diatribe makes ample reference to the work of Dr. Keith Green at the Medical College of Georgia and insinuates that his research demonstrates SLS's lethal properties. Dr. Green has disassociated himself from this nonsense. He explains that all he did was investigate the substance's irritant potential by putting it into the eyes or rabbits. (While such tests are disturbing, they're sometimes necessary.) Green did, in fact, detect some SLS in the test animals' tissues, but he observed no toxic effects. It's not surprising — since SLS is used in toothpaste, it has undergone a battery of toxicity tests, including ingestion experiments. Scientists conducting those tests encountered no problems, even though SLS remains in the bloodstream for about five days.

While it does cause the rare skin or eye irritation, SLS is a remarkably safe substance. A few years ago a team of researchers in England caused a stir by implying that detergent residues on improperly rinsed dishes could lead to irritation of the gastrointestinal tract. The press had a field day with the story: "Danger lurks in dishwashing liquids!" screamed the headlines. The researchers were compelled to point out that their study involved only six rats that had ingested detergents with their drinking water at doses thousands of times greater than those any human would be exposed to.

There is one real, and bitter, problem associated with sodium lauryl sulfate. If you brush your teeth with an SLS-containing toothpaste and then drink orange juice, you may not like the taste. Oranges contain a fair bit of citric acid, which has both a bitter and a sour taste; for some reason, sodium lauryl sulfate enhances the bitter taste and leaves the sour taste unaffected.

But even this effect doesn't happen to everyone. You have to possess the genetic predisposition.

While we mainly use SLS for cleaning, the substance could have another interesting application in the future. Researchers have discovered that SLS resembles a compound called paradaxin, which is found in Red Sea flatfish. Paradaxin is known to repel sharks. With this in mind, scientists tested SLS as a repellent for lemon sharks, a species that sometimes attacks humans. It worked. So, the next time you plan to swim in shark-infested waters, have some shampoo at the ready. Now if only we could find a way to repel the Internet sharks and prevent them from taking a bite out of the gullible. The Internet can be a wonderful resource, but it can also be a frightening quagmire. Internet technology coupled with ignorance of chemistry and toxicology has resulted in a misguided attack on sodium lauryl sulfate. The truly alarming aspect of this entire issue, however, is how a handful of pseudoscientific arguments can cause virtual panic among consumers. As you can see, I find such irresponsible innuendo spreading very irritating. It makes me want to go and relax in a nice, sodium-lauryl-sulfate-laced bubble bath.

pHooey to pHake Health Claims

We now know the cause of all degenerative diseases. I have learned of this dramatic breakthrough in a glossy publication that came through the mail. It's called *The Vaxa Journal*, and at first glance it appeared to be a scientific magazine, but it is actually a vehicle for marketing a variety of miraculous-sounding supplements. One cannot afford not to take "Buffer-pH," we are told, because "adding ten years to life is as easy as following the directions on the label!"

I must admit that I was intrigued as I flipped through the journal and encountered repeated references to pH — here was something a chemist could really sink his teeth into. Scientists routinely use the pH scale to measure the acidity of a solution, be it blood, wine, or tap water; but I never realized that the wrong pH could be killing us. Yet that is exactly what the *Vaxa Journal* contends. "The pH Factor: The Real Silent Killer," reads the headline. The solution to longevity is simple: pop a Buffer-pH capsule daily and lower your risk of heart disease, stroke, cancer, obesity, multiple sclerosis, and Alzheimer's disease.

How can this be? Is it possible that research scientists around the world have somehow missed this simple answer to the mystery of degenerative disease? The *Vaxa Journal* articles sure seem to make a good case for it. They examine in detail how blood pH is critical to maintaining good health, and they describe the calamities that ensue when pH ventures outside the normal range of 7.35–7.45. Physicians, of course, know all about this. They know that acidosis can cause weakness, nausea, confusion, and even death. They know that alkalosis, a condition in which the blood is too basic, can cause irritability, muscle twitching, and convulsions. They know that these conditions require immediate treatment. But they also know how rarely they occur.

The human body is a remarkable machine. It relies on a variety of safeguards to keep blood pH constant. Our blood constitutes a buffer system — meaning, it has components that can react with excess base or excess acid. Carbon dioxide, which is produced by the metabolism of food, dissolves in blood to produce carbonic acid, and carbonic acid can neutralize any excess base. The bicarbonate ion, also present in blood, will promptly take care of any surplus acid. The level of carbon dioxide in the blood adjusts to a body's rate of respiration. If blood pH drops — which actually means that the blood has become more acidic — we breathe faster, exhale carbon dioxide, and thereby reduce the acidity. If the pH rises, our respiration is inhibited, we exhale less carbon dioxide, and again our pH returns to normal. Then there is a backup system. The kidneys regulate blood pH by excreting or retaining acid.

This regulatory system is highly effective and can meet most challenges, but not all. Conditions that affect respiration, such as pneumonia, injury to the brain's respiratory center, or morphine overdose, may lead to respiratory acidosis. Uncontrolled diabetes, kidney failure, and starvation may all result in a decrease of blood pH, causing metabolic acidosis. This will also occur if a person overdoses on substances like alcohol or antifreeze, which the body metabolizes into acidic compounds. Respiratory alkalosis is usually caused by excessive loss of carbon dioxide due to hyperventilation. That's why some people who have hysteria or panic attacks experience physical reactions. The treatment for these reactions is having the attack victim breathe into a paper bag — thereby reinhaling the exhaled carbon dioxide. Metabolic alkalosis can be caused by loss of hydrochloric acid from the stomach through persistent vomiting, kidney disease, or excessive intake of antacids.

These ideas were elegantly introduced to the general public in 1969 by Michael Crichton. Long before he terrified us with carnivorous velociraptors in *Jurassic Park*, Crichton scared us with extraterrestrial viruses in *The Andromeda Strain*. Written in an era of space-exploration fever, the novel explores the possibility of a virus from outer space invading the earth by hitching a ride aboard a satellite that crashes near a small American town. The resulting infection wipes out all the local inhabitants, save for the town drunk and a crying baby. Crichton is a physician with a sound scientific background, which he ingeniously draws upon in weaving a plausible plot. The invading virus, we learn, can only multiply in human blood that has exactly the right pH. And guess what pH our invading virus prefers? That's right — 7.4, the pH of normal blood.

But why does the baby survive? Because the infant cries and screams at full throttle, thereby exhaling a lot of carbon dioxide. This, as we have seen, raises the pH. The drunk, on the other hand, runs out of wine and is obliged to drink Sterno, the jellied fuel that comes in a little can and when lit warms buffet dishes from beneath with a nearly invisible blue flame. The active combustible ingredient is methanol, also known as wood alcohol. The effect of drinking methanol is similar to the effect of drinking ethanol, the intoxicating ingredient in the liquors we drink. But there is a crucial difference: methanol is much more toxic. When methanol is metabolized in the liver, it is converted to formic acid. This substance acidifies the blood, alters the pH, and wreaks havoc. The town drunk was lucky enough to have consumed just enough methanol to change the pH of his blood slightly; there was enough in his system to kill the deadly virus, but not enough to kill him. Science fiction to be sure, but certainly relevant to daily life.

Now let's get back to the science fiction that is not relevant to daily life. Like the notion that all degenerative disease is triggered by "unhealthy pH." The writers of the remarkable articles in *The Vaxa Journal* discuss the potential dangers of acidosis but then go on to insinuate that this is a common problem. Why? Because the North American diet, which contains a lot of protein, leads to an "acid residue" in the body, the "seed-bed of all degenerative diseases." While it is true that certain foods can have an acid or basic residue, the fact is that this is irrelevant in face of the healthy body's buffering ability. If there is too much acid, the body neutralizes or excretes it. The pH of the urine may be affected, but not the pH of the blood. The journal makes one last fantastic allegation: cholesterol deposits in our arteries are the result of low pH. Diagrams published with the articles show progressive blockage with a drop in pH. It looks very dramatic, but there is absolutely no scientific evidence for this.

Still, I'm sure that many people have swallowed the Buffer-pH capsules and the pseudoscientific lingo that goes along with them. What do the capsules contain? Small amounts of antacids, which aren't enough to deal with heartburn, let alone change the pH of blood. Such nonsense is enough to make a grown scientist cry just like the baby in *The Andromeda Strain*. Of course, the *Vaxa* people would say that this is a good response because it raises blood pH and keeps one healthy. I'd say that's pure science fiction. And I'd also say that we have had just about enough of all this silly stuff.

ACKNOWLEDGMENTS

Coaxing a genie out of a bottle is not an easy task. It requires some help. Colleagues Ariel Fenster and David Harpp, with whom I have worked for over twenty years, made many valuable suggestions. Professor Harpp read every word in the manuscript and offered numerous helpful comments. But more than that, his dedication to teaching and his drive for excellence has been an inspiration throughout my career. Thanks are also due to Professor Arthur Perlin, who so many years ago, as my thesis advisor, taught me how to pursue a problem in a proper scientific fashion. Dr. Seymour Mishkin graciously answered many of the medical questions that arose during the preparation of the manuscript.

I am also grateful to the many students and members of the public whose probing questions over the years stimulated me to investigate some of the most fascinating aspects of chemistry. It has been a pleasure to work with a number of excellent colleagues at the Montreal *Gazette*, the Discovery Channel, and CJAD radio in Montreal. They have helped make interacting with the public a memorable and enjoyable experience. Rob Braide, general manager of CJAD, has been particularly supportive, while thanks also go to Melanie King and Tommy Schnurmacher, radio hosts who have contributed wit and substance to our programs over the years. Robert Lecker, my publisher, deserves my gratitude for suggesting I undertake the writing first of *Radar, Hula Hoops and Playful Pigs*, and now of *Genie*.

Last, but not least, writing a book requires family support. My wife Alice, and daughters Lisa, Debbie, and Rachel have had to fend for themselves many a Saturday and Sunday morning while I was conjuring up *Genie*. Nary a complaint. That's greatly appreciated because, as you know, once the genie is out of the bottle, you can't put it back. There are always more stories to be told. *That's the Way the Cookie Crumbles*, the third book in this series, is already in the works.

INDEX